Successful Professional Gamblers

Lance D. Williams

Published by Lance D. Williams, 2023.

While every precaution has been taken in the preparation of this book, the publisher assumes no responsibility for errors or omissions, or for damages resulting from the use of the information contained herein.

SUCCESSFUL PROFESSIONAL GAMBLERS

First edition. August 13, 2023.

Copyright © 2023 Lance D. Williams.

ISBN: 979-8223186809

Written by Lance D. Williams.

Table of Contents

Introduction

For a few years now, I've thought about making a lottery book to share with people how I've figured out the lottery and won. Once you learn it and can actually see why the numbers come out the way they do, it is fascinating. I want people to see that there is a logical way to win if you're willing to put in the work to learn about the games you play. There are a lot of people who think it's impossible to win or figure out the lottery because it's "random." I used to be one of those people who refused to gamble because I thought it was a waste of money and not worth it.

When I started studying the lottery and playing it, I eventually became successful at it. I never would've imagined that I'd become a professional lottery player, but the results have been great, and I'm successful doing it. One of the reasons I've put off doing a lottery book is that I wasn't sure I wanted to share my methods and secrets with people about how I win. I've shown and taught people close to me how I do it. But I've also been willing to tell people the basics when they ask me about the lottery. Sometimes someone will ask me if I win by playing the lottery. I'm honest and say yes, and I tell them the basics. At this point, I want to share with anyone who reads this book some of the things I've learned. I want others to learn and see for themselves that it's possible.

Gambling is becoming more and more popular, especially with sports betting. I'm strictly a lottery player; I do what I know best. But I'm always curious about other professional gamblers and learning how they've become successful with the things they do. In the past few years, I wanted to find people who were willing to share their stories with me about their gambling experiences. In this book, you learn about various people who gamble in various ways. It's interesting learning about the methods they use and getting to the point where they became successful. One of the common things I noticed with many of us is that we observed and studied things and used that to our advantage to win. Paying attention to the details is important. Learning about gambling is important before playing.

I hope anyone who gambles are smart with their decisions. A lot of people lose money. I hope that by publishing this book, people will be able to learn helpful tips to succeed. But also learn what not to do to minimize failure. Winning is **<u>NOT</u>** guaranteed. When you gamble long enough, you're going to experience many losses. Try to avoid losing all your money in the process. Learn for yourself but learn from others. What I mean by that is that, whether you read these stories or information from other sources, you can learn from our strategies. But it's important to learn on your own to understand what you're doing and find what works for you. Just because I'm successful with the way I do the lottery doesn't mean you'll easily pick it up and start winning too. It took me a while to learn and understand how the lottery system works. I didn't start winning until I learned how the system worked and knowing what to look for.

DISCLAIMER: Most of the other stories in the book have been edited for clarity. Some financial numbers were verified. For those dealing with a **gambling addiction** or those concerned about someone with a **gambling problem,** please receive guidance on treatment options and support groups.

My Lottery Experience

I was sitting in my office on March 28, 2016. I was thinking about my schedule for the next few months, after I'm finished with my last assignment for a client in April. I'd probably have some free time for a while; I didn't have anyone else to do work for. Then I thought about the Powerball. In January 2016, the jackpot had reached over a billion dollars. My girlfriend at the time wanted to buy some tickets, and she asked me to give her some numbers to play. I gave her 6-9-40-42-56 and 19. I also gave her 1-20-48-53-58 and 11. I didn't take it seriously; my thoughts were that people were wasting their money playing the lottery. I was an entrepreneur, so I always said, I'd rather invest in building a business than spend on random numbers. She played and won a few dollars. The winning numbers were 4-8-19-27-34 and Powerball 10. People from three different states won the jackpot and split the $1.6 billion.

For some reason, I was thinking about when she played. I realized I was being ignorant about the lottery. I was judging something without really trying to understand it. Growing up, my mom, dad, and grandmother played the lottery. Living in Toledo, Ohio, we're right on the border with Michigan. So, my mom and dad would both drive to Michigan to play their lottery and played the Ohio Lottery. When my grandmother wasn't home, she wanted me to write the lottery numbers down when it came on TV. I remember she had a notebook where she wrote the Pick 3 numbers that came out. My mom buys a bunch of lottery books that tell you what the best numbers are to play. I always thought she wasted money on those scams. How could they know what numbers might come out if the lottery is random? Especially months in advance. Even fortune cookies often have numbers that are "lucky."

I was thinking about all these things. I always like to give myself challenges and goals. That's when I decided my new challenge was going to be trying to win the lottery. Not by guessing, not by auto picks, but actually figuring out what numbers will come out. Like I mentioned before, my grandmother would write down the numbers that came out. But that's all she did. There wasn't any detailed information she was using to try to figure out what numbers to play.

On April 16, 2016, I decided I was going to write down every number that came out in 2015, so I had a database of information to work from. I did all the Ohio Lottery games: Pick 3 midday and evening. Pick 4 midday and evening. Pick 5 midday and evening. Rolling Cash 5, and Classic Lotto. I also did the national lotteries, Powerball and Mega Millions. Eventually, I added Lucky for Life as well. So, I put in a lot of work tallying up every number that came out the previous year. I was winging it. I had no idea what I was doing or why I decided to do it this way. If I was going to play, I wanted to take it seriously. I was treating the lottery as an investment. If you're investing in a business, you want to make sure you have as much information as possible before investing. If you're buying a vehicle, you want to make sure you have as much information as possible and test drive it before you pay for it. The same thing is true with the lottery, if you're going to invest time and money into it. Give yourself the best chance to win, so you're not wasting time and money.

After I spent a few days writing down all the numbers, now I had to figure out what to do with this information. Like I said, I was winging it. There was going to be a lot of trial and error trying to figure out what numbers to play. One of the first things I thought might be key information was looking at what numbers came out the most and what numbers came out the least. I studied different combinations of numbers in all the games.

On April 20, 2016, I finally wanted to start playing the lottery. I think I played a few Rolling Cash 5 combinations. I also played Classic Lotto. I think I had one Classic Lotto number, and I don't think I had any Rolling Cash 5 numbers. When all the numbers came out, I tallied up the numbers to add to my database of notes. I always wanted it to be up-to-date and accurate. The next day, I wanted to try midday Pick 3 and 4. Trying to figure out those numbers to play was harder than the Rolling Cash 5 and Classic Lotto. Rolling Cash 5 is 1-39. Classic Lotto is 1-49. Pick 3 and Pick 4 are 0-9. But it could be different numbers or the same numbers. I found that to be more challenging. I was trying to stick to facts and figure out the information I had in my notes. But I was tempted to do what most people do—guess or something random like a birthday. I thought about playing my office number, 125. My parking spot was right outside of my office window. So, I look outside the window sometimes. There's a woman who works in the office building across the parking lot; she usually parks next to me. I was looking at her license plate and thought about playing that for Pick 4, because it had some of the same numbers. I don't remember what the exact numbers were, but I know it had at least 66 in it. I decided to stick with focusing on my notes. I don't remember the numbers I played, but I didn't win. But her license plate number actually came out. Depending on how I played it, I could have won $1,198 or $5,000. I couldn't believe I missed out on winning.

Within the first month, I realized I was seeing a pattern with the Rolling Cash 5 numbers. Some of the numbers went together often. That became my go-to game to play. I started playing different combinations daily, and the lady that worked at the gas station down the street gave me the nickname "Rolling Cash." One of the things I noticed about the Ohio Lottery games was that when one of the numbers came out, usually the next day at least one of them would repeat again.

I started to focus on what number I thought had the best chance to come out again the next day and figure out what numbers would go with it. So, for example, with Pick 3, if 123 came out and I thought number 2 would repeat the next day, I worked off the information from number 2. I noticed with Rolling Cash 5, there was a pattern that would happen often. RC5 starts Sunday and ends Saturday, then it basically starts over. I would usually know by Wednesday how the pattern was going to go. I just had to figure out what number combinations would go together. The numbers are 1-39. A lot of times, by the end of the week, most of the blocks of numbers will end up being used, and it changes to other blocks throughout the month. So, let's say in one week 1, 2, 3, 5, 7, 8, and 9 will all eventually be used. Then the next week it might be 20, 22, 23, 25, 26, 29. Then maybe the week after that, it could be 13, 14, 15, 16, 17, 18, and 19 that eventually will be used. Once you figure out the pattern, half the challenge is done. Now you can focus on what combination of numbers goes together. A real example: 5-15-32-33-37 comes out. In my notes I have how many times numbers have come out different times throughout the years. **5 (219), 15 (232), 32 (220), 33 (233), 37 (231).** So, you can see why these numbers fit together. It doesn't always go like that, but it happens often.

Classic Lotto is different from the other Ohio Lottery games. Those are daily Sunday-Saturday. This one is just Monday, Wednesday, and Saturday. It also has the largest payout. It starts at $1 million and continues to go up until someone wins. I've seen it get up to like $18 million over a couple of years until there's a winner. So, it's the hardest one. The numbers are 1–49. Figuring out the pattern is harder since there are fewer days and you need to get six numbers for the jackpot. One pattern that I did notice that makes it easier to win is that usually every month it would have a bunch of numbers in a row. So, for example, 1-34-35-36-37-49. Or sometimes it's a couple numbers together like 11-23-24-32-40-41. If you're able to figure out when it will happen and what numbers are going together, then you have a good shot of winning at least $70-$1,500 if you don't get the jackpot.

I'm a very disciplined lottery player. So, I rarely spend over $5 a day. I probably average around $3. I just play when I'm confident that I've figured out certain numbers that will come out. In the first six months, I might've spent $10 playing a bunch of different numbers in one day.

Once, I spent $18 in one day and won $10 back. A lot of times, with Rolling Cash 5, I would win my $1 back. Sometimes I would win $10. I had to learn a lot of lessons and break bad habits.

Early on, I would play the numbers I wanted to come out instead of focusing on how the system wanted the numbers to come out. I would feel like certain numbers were due to come out, and that's not always the case. The more I followed the numbers and spent hours studying the trends, the better I got at learning what numbers go together. That's when I started winning $300, meaning I got 4 out of 5 numbers with Rolling Cash 5. I would win $70 with Classic Lotto, meaning I got 4 out of 6 numbers. With Pick 3, sometimes I would win between $41 and $500, depending on how I played the numbers.

Those were the games I studied and played the most. I rarely played the other games; I couldn't figure out the patterns and didn't feel like I would win. I still collected data for a few years just to have it. Eventually, it became overwhelming trying to keep track of 11 different games. So, I dropped Pick 4, Pick 5, and Powerball. I continued to keep track of Pick 3, Rolling Cash 5, Classic Lotto, Mega Millions, and Lucky for Life. It's still a lot to keep track of, but those are games I've been able to figure out and win.

When it comes to the bigger games, Rolling Cash 5, Classic Lotto, Mega Millions, and Lucky for Life, my goal wasn't even the jackpots. I could make a good living by consistently winning some of the numbers. So, with Rolling Cash 5, I'm fine with $300 if I don't get the six-figure jackpot. With Classic Lotto, I'm fine with $70 if I don't get the millions in the jackpot. With Mega Millions, I'm fine with $200 or $500 if I don't get the millions in the jackpot. With Lucky for Life, I'm fine with $150 or $200 if I don't get the six-figure or millions for the top prizes. Plus, with Pick 3, there's plenty of opportunities to make life-changing money. It all adds up. Like I tell people all the time, there's nothing wrong with being Ichiro Suzuki and getting singles, doubles, and triples; you don't have to be Barry Bonds and hit home runs every time.

Yes, winning a billion-dollar jackpot would be amazing, but I'm just as happy when I win $500. For me, it's not even about the money; I was already fine before I played the lottery. The best satisfaction is knowing that every time I win the lottery, it's because I actually figured it out. Like in school, when the math teacher would say, "Show your work." I can show my work and how I figured out what numbers would come out.

I'm currently in year seven, playing the lottery almost every day. I have 8 years of notes on most of the games I play. It gets exhausting keeping up with all of the games, and at some point, I'll either reduce the games I play or retire doing all of them. There are times I want to hire someone to keep track of the games for me, but I don't trust anyone to do it. When I do it, I know that it's accurate. All it takes is one mistake to mess up seven years of notes. It's not necessarily drastic, but in Rolling Cash 5, for example, there's a big gap between getting 4 numbers and winning $300 and getting 5 numbers and winning $300,000. It's always funny when I tell people that playing the lottery is one of the primary things I do. Their reaction is like I had in January 2016, that the lottery is random, impossible, wasting money, etc. Judging something they don't realize is a legitimate way to make money, if you're smart with the way you play.

Most people aren't going to take time to research all the numbers and build a database of information. Taking time to study games for hours on a daily basis and seeing how the trends work. Or have the patience to learn through trial and error. It might take years until you figure out how the lottery games work. It could take a long time to win. There might be times you get lucky with it. But actually, figuring it out on your own, where you can show your work, takes time. For most people, it's not going to be worth it. I don't think there's a better way to make money with profit returns.

I'm just going to give an outlier example. Let's say you decided to follow the lottery every day, building up a database of notes. It doesn't take any money, just time. It takes two minutes to write down the numbers that came out. For some unfortunate reason, you're absolutely broke. You have no money in your pocket. You're walking down the street, and you find 50 cents at 11 a.m. You want to turn that 50 cents into more money, so you look at your lottery notes, and you have a good feeling 111 is going to be the numbers for midday Pick 3 based on the information in your notes. You play it at 12:15 p.m. Then, by 1 p.m., you find out that it came out. Now that 50 cents turns into $250, you cash the ticket in at the store. You go back to your notes to look at the evening games. You play Rolling Cash 5 and win $300. Now you have $549. You decide at 8:30 p.m. to play Mega Millions after looking at your notes. You end up getting 5 out of 6 numbers and winning a million dollars. What else can you do where you start the day with absolutely no money and find a way to make a million dollars 12 hours later?

Obviously, most of the time, it's not going to work out perfectly like that, but trust me, it's absolutely possible if you put yourself in a position to win. Even if you take away two of those wins, it's $250. You did that with maybe 30 minutes of work looking at your notes and trying to figure out what numbers would come out.

A lot of times, when the Mega Millions and Powerball start to get into the $500 million to billion-dollar range, that's when more people are interested in playing. Sometimes the office buys a bunch of tickets together. If you're a casual player, wouldn't you want to give yourself the best chance to win? Even if you just took notes on all the Mega Millions numbers on Tuesday and Friday, when it reaches a point where it piques your interest to play, you can look at your notes, have the most information possible, and turn $2 or however much you decide to spend and make a profit.

People have no problem investing in stocks, going to the casino, doing sports gambling, or getting involved with crypto. I have friends that do those things, but when I mention the lottery, that's when they act like I'm the one wasting money. But the best investment, in my opinion, is the lottery if you do it right. There's been plenty of times I've won the lottery without using my own money. I find change on the ground all the time. People don't even know they're helping me make a lot of money. I pick up pennies off the ground; it goes with the change I use for the lottery sometimes. That 50-cent example I used earlier is something I've personally done. I wasn't broke at the beginning of the day, but I was able to make more money each time, and it started with just 50 cents that I found on the ground.

I've been successful playing the lottery. I'm considered an expert in some games. I've been asked to teach lottery classes at colleges. The lottery has changed my life. I've been able to semi-retire in my 30s. I pick and choose when I want to work. I'm still active with a lot of stuff. I've been able to help people with lottery winnings. I'm not guaranteeing that people are going to win. That depends on each individual; I just want to encourage people to have this option available even if they don't play daily. Give yourself the best chance to win if you're going to do it. I see so many people play the lottery, and they spend so much time playing a bunch of random numbers. It's better to take time to figure out what numbers might come out and play them.

I would win a lot more if I played all the options when I narrowed it down. But I pick what I'm most confident with and go with it. Like an NCAA bracket, I do one and roll with it. I've picked the correct winner the past four years and had most of the Final Four picks correct. I want to win because I figured it out, not after several guesses. Through my years of playing and learning, I've figured out so many ways to make money. There are a lot of tips and tricks you can use once you figure out the basics. I would say to start with basic goals. Try figuring out what two numbers are going to be and building from there.

Like I said, you don't have to always go for the home run (the jackpot). You also don't need to start out following every game like I did. Pick one that you feel the most confident you can win and master that first before moving on to other games. You don't want to overwhelm yourself. It doesn't take a mathematical genius to win. You don't have to do it like Jerry and Marge Selbee. It just takes accurate notetaking, lots of studying the notes to learn the trends, and patience. It might take months, it might take years, and learning and winning might not happen at all, depending on who you are.

Don't spend recklessly; be wise in how you invest your lottery money. If you only spend $3 a week, that's fine. You don't have to spend $30 a day playing a bunch of different combinations and hoping one of the tickets wins. Pick the numbers that make the most sense and go with it. If you don't win, oh well, focus on the next time you play.

I'll end with basic examples. To give you an idea of how I do it. Let's say you're playing Pick 3 this evening. You're looking at your notes. The previous night, 702 came out. If you feel like 2 is the number going to repeat, work off the information from 2. So, if you're looking at your different pages of notes, I've been doing the lottery since 2016, so I have 7-8 years of data on how many times each number has come out. I'm looking for numbers that match well with the number 2, based on my notes. Let's say I'm looking at the page to see how many times the numbers have come out in the past year. #2 has 161 times. Then I'm going to look at numbers that might match it. If #1 has 162 times and #5 has 163 times, then there's a good chance 215 could come out in some type of order.

I have additional notes that help me figure out whether I should box it or play it straight. Sometimes I do both. But let's say I decide to play 125, box it, and just do 50 cents. If 251 comes out, then I would win $41.50. That's a real example, with the actual numbers at that time. It doesn't always work out perfectly matched like that, but a lot of times it's two that will go together, and it might take a little more work figuring out what the third number will be.

Hopefully, I made that simple enough to understand. But that's basically how it works. Keeping track of the numbers that come out, trying to find the number that has a chance to repeat, and finding out numbers that could go well with it for that day.

I could write a whole book talking about the things I've learned and showing the exact ways I do it for all the games. When I show people, they're amazed at how possible it really is to learn the lottery and make money doing it. I read articles all the time of lottery winners talking about how they won. I'm just curious to see what they say. Honestly, it always seems silly and lacks substance. I just read an article last night about how impossible it is to be a professional lottery player. Learn for yourself, don't just believe what other people tell you. I tell people all the time. Just follow it for a month. Write down the numbers every day and you'll see different patterns and notice numbers that come out together throughout the month. I want people to visualize and understand how a number or numbers repeat frequently and that can be used as a starting point. You can look at every lottery site and see how many people have won that day and how much has been paid out. So many people win every day.

People only focus on the jackpot as winning. But as long as you win more than you spent, it's a win. If you spent $2 and won $25. That's a win! I also want to finish off by saying that this is how it works for the Ohio Lottery with some examples I used. I haven't had an opportunity to start learning about other state lotteries and how the trends work. I'm pretty sure it's basically the same. I've seen similar trends with the national lotteries as well.

A real <u>Lucky for Life</u> example: I won't give out the numbers that came out. I'm not going to give all the answers away, I want people to do their own work. But I'll show how many times those numbers came out so you can see why the combinations went together. (**Sunday**):

(144), (126), (141), (159), (138), LB- (73). (Monday): (145), (143), (146), (129), (142), LB- (75). This is a good beginning of the week example, you're able to see what numbers to look for to fill in the gaps for the rest of the week. So, for the Lucky Ball looking for one that came out 72 times or 73, 74, 75, or 76 is likely to happen. Same with the other numbers. A real <u>Mega Millions</u> example: Again, I won't give out the numbers that came out, just how many times. (Tuesday): (52), (61), (52), (53), (48), MB- (27). (Friday): (54), (50), (53), (49), (44), MB- (26). So, this is a good example of most of the numbers fitting together. It's not always like this, but at least you're able to see, if you have accurate notes, you give yourself a great chance to win money if you can find the right combinations.

A real <u>Powerball</u> example: Once again, not giving the numbers that came out, just how many times they came out in my notes. I recently started over on the Powerball, so my years of notes aren't as long as the other games. This was on (Monday): (27), (20), (20), (20), (26), PB- (15). This was on (Wednesday): (13), (17), (12), (21), (24), PB- (9). This was on (Saturday): (20), (22), (23), (26), (26), PB- (14). So, during this week a lot of the numbers that came out were around the same amount of times. These are basic real examples just to show how combinations fit together. But I have more notes and information I use in various games to help me figure out numbers. I don't want to give out all the goodies.

Here is an Ohio <u>Pick 5</u> midday example: So, this is how many times the numbers have come out on Friday. (22), (24), (23), (23), (23). This is how many times the numbers came out throughout the year so far. (169), (171), (171), (170), (171). It doesn't always fit perfectly like these examples, but it happens often.

Earlier I gave a quick example on Rolling Cash 5, Classic Lotto, and Lucky for Life. Here are some full week examples so you can get a better visual:

<u>Rolling Cash 5</u>:
   (Sunday): **(282)**, **(280)**, **(289)**, **(238)**, **(274)**.
   (Monday): **(280)**, **(279)**, **(267)**, **(294)**, **(275)**.
   (Tuesday): **(279)**, **(268)**, **(249)**, **(282)**, **(297)**.
   (Wednesday): **(320)**, **(268)**, **(268)**, **(282)**, **(274)**.
   (Thursday): **(251)**, **(321)**, **(277)**, **(283)**, **(255)**.
   (Friday): **(286)**, **(283)**, **(281)**, **(286)**, **(264)**.
   (Saturday): **(284)**, **(282)**, **(269)**, **(287)**, **(256)**.

<u>Classic Lotto</u>:
   (Monday): **(127)**, **(142)**, **(139)**, **(123)**, **(145)**, **(158)**.
   (Wednesday): **(119)**, **(130)**, **(120)**, **(128)**, **(135)**, **(146)**.
   (Saturday): **(139)**, **(145)**, **(136)**, **(132)**, **(133)**, **(119)**.

<u>Lucky for Life</u>:
   (Sunday): **(188)**, **(172)**, **(146)**, **(159)**, **(187)**, **LB-** **(98)**.
   (Monday): **(189)**, **(180)**, **(147)**, **(176)**, **(177)**, **LB-** **(103)**.
   (Tuesday): **(180)**, **(179)**, **(160)**, **(177)**, **(178)**, **LB-** **(99)**.
   (Wednesday): **(194)**, **(166)**, **(166)**, **(159)**, **(174)**, **LB-** **(99)**.
   (Thursday): **(158)**, **(178)**, **(167)**, **(180)**, **(175)**, **LB-** **(100)**.
   (Friday): **(172)**, **(172)**, **(190)**, **(179)**, **(168)**, **LB-** **(92)**.
   (Saturday): **(173)**, **(167)**, **(161)**, **(178)**, **(188)**, **LB-** **(93)**.

Wherever you are, learn how the games work before you play. My advice to get started, add up every number on whatever game you're playing. If you're starting in 2024, go back and keep track of every number that came out in 2023. I would even do 2022 as well. Yes, that requires effort. A lot of people just want this to be easy. It gets easier once you get the hard work done first. Make sure you don't miscount. Double-check what you're doing. If the baby starts crying, make sure you know where you left off, otherwise you're wasting your time with wrong information. This at least gives you a decent size database of information to follow.

Some people will decide to be lazy and use a website that tells you how many times numbers have come out. When you do it yourself, that's when you pick up on a lot of helpful information about each number daily, weekly, monthly, and yearly. You'll learn what number came out the most. What came out the least. You'll most likely see that most numbers that come out are close to even with how many times they come out. You learn what numbers go together the most and learn trends and patterns. It's like looking at *Where's Waldo?* books, where you look and examine things long enough, more things will stand out.

When people say it's random, no. There's a method to the system. That's why combinations come out the way they do. Continue to take accurate notes from there. Every time numbers come out, add on. For example, if #1 came out 10 times, and today it came out again, now it's 11. Examine everything you can. Numbers that repeat. Numbers that go together often, etc.... There are so many things to look for once you get used to studying and understanding trends.

With Pick 3 midday for example, if #0 has come out 103 times in 2023 so far and #1 has come out 100 times so far, and #1 might be your repeat number today. I have a good track record of knowing when the trends are due to come out, because of my experience of following it for so long. So, this could be a good time to go with 111. So, #1 will catch up with #0 and both be at 103 times. That's not guaranteed. I'm just giving you an idea, that keeping track of numbers, learning trends, and looking at every possibility for certain numbers to go together, will help you win often. Add whatever information you need; you might pick up on certain things you noticed. I have pages of notes for each day.

So, Sunday-Saturday on the daily games, I keep track of how many times numbers come out that day. That way you know what number comes out the most that day, the least, what might be due, there's so much you can learn by tracking all the numbers that come out. Going back to the Pick 3 example, even keeping track of the 1st, 2nd, and 3rd number that come out each day. So, if you're looking at your Wednesday notes and your page for the 1st number that day has #5 that came out 4 times and #6 came out 3 so far. It might be time for #6 to be the first number. It doesn't mean it will happen, it's just a data point to look at. Doing this helps me figure out when I should play certain numbers straight or box it. I've been doing this for years, so I know what to look for.

Most people won't pick up on a lot of these tips until you study it yourself and actually see it when you're going over your notes. Here's something else I study. I'll give you real examples on this one. Ohio Pick 3 midday in 2023 so far: Double numbers that come out like 1<u>01</u>, I always underline. **January (9), February (3), March (5), (2) triples** like 111. **April (7), May (7), June (9), (3) triples. July (7), (1) triple. August (12), (1) triple.** Ohio evening in 2023 so far: **January (10), February (4), March (9), (1) same # twice <u>99</u>4. April (10), May (6), (1) triple. June (8), (1) same # twice 854. July (10), (1) same # twice <u>22</u>0. August (5), (1) same # twice 417.** Learning and studying everything has always been helpful and I can apply to figuring out numbers eventually. Same thing with games like Mega Millions, Powerball, etc.

Keeping track of the numbers for each day will help you find the most popular, the least popular, when it might be time for 30 to catch up to 31 to be even with how many times it came out. By the end of the year on all the games, you'll see that it's close to even with how many times numbers come out. The end of the year number is what I continue to use with my totals. So, with game _____ if #4 came out 50 times in 2022. 42 times in 2023. I'm starting my 2024 notes with #4 (92) and continue to add to the number. That's what helps you know what combinations to go with. If #4 is (92) and #9 is also (92) they'll go together at some point when the time is right for it.

Games like Rolling Cash 5 and Classic Lotto I always add up the numbers that come out. So, RC5 for example on 1/1/23: If it's 1-2-3-4-5 the total is (15). Then on 1/4/23: If 1-2-3-4-6 comes out the total is (16). You'll see weekly or during a specific month that many of the numbers that come out, the totals will be around the same. That's also helpful information to know during the year. Let's say 1/1/23: The total is (15), there's a good possibility on the 1st in one of the other months a total could be 14, 15, 16. You just have to figure out the combinations that will add up to whatever number on that date you're looking at. Again, that's not guaranteed, it's just another thing I've observed happening frequently over the years.

Another thing I do as well, I underline all the numbers that go together each time. You'll see that numbers repeat together often. So, if 1-2-3-4-5 comes out 1/1: Then 2-3-10-15-20 come out 1/15: I underline 2 and 3. Many numbers go together throughout the month, I would say at least 10 times in a month you'll have all the numbers underlined. Meaning numbers that went together often that month. For example, 2 and 3. 16 and 36. 3 and 36. Things like that. Always look at numbers that come out together often, it will help you figure various things out when looking at your notes. That is what I mean, when I say **EXAMINE EVERYTHING**, there are so many things you'll pick on if you're studying detailed notes for hours every day.

Most people won't take the time to learn, but that's what will separate you from others and help you win consistently, when you know what to look for, and pick up on every pattern and trend. Some people that teach their lottery systems make it out to be complicated mathematical probabilities to win. I believe in KISS "Keep it simple, stupid!" If you have the patience and ability to keep track of numbers every day, without making mistakes, eventually you should be able to pick up on things that will help you win. If other books, articles, videos, or other sources give you generic methods, promising they have specific answers to win, it's probably wrong. Every game is different, there are certain things you can apply like the examples I've given you. But ultimately, it comes down to you figuring out what to play.

From my experience you must know what to look for to win. So, every week the lottery basically starts over. There might be new trends and combinations to look for. If a new game starts Sunday, maybe wait until Monday or Tuesday to play, so you have some information to use for that week. Like I gave you in the Lucky for Life example. This is important to remember, focus on daily and weekly patterns and combinations. Like the examples I gave you with Rolling Cash 5 and Classic Lotto, knowing when blocks of numbers will be used or when a bunch of numbers will come out in a row. Anything beyond that is useless in my opinion. If someone claims they can predict far in advance what the numbers will be, needs to prove exactly how they came to that conclusion by showing their work for me to believe it. Using information close to when the lottery is due is the most common sense thing, so you're able to see where the current pattern or trend is that week.

Just because a mathematics professor says, "You're more likely to get struck by lightning than win the lottery." Doesn't mean you have to believe it. Too many people focus on reasons you can't win, instead focusing on what you should do to give yourself the best chance to win.

If you wanted to do a puzzle, most people probably wouldn't say I have a 1 in 300 million chance to complete it. You dump out the pieces and start trying to put it together. Same with the lottery, you use the information you have in front of you, and you start putting the combinations together. The only lottery system is the system itself. The facts are there for you to use. Focus on how many times numbers have come out. That's the main key to figuring out combinations. But also use other information to your advantage as well, like the tips I gave you earlier about examine everything, and you'll continue to find pieces to the lottery puzzle. I'm going to keep repeating those points. I'm not telling you anything you can't learn as well. Just follow the lottery every day for years like I do. I could easily sell my years of notes for thousands of dollars, but it's not about making money from as many people as possible. In most of my books I encourage people to learn new things they might not know about. If you're really interested, you'll make the commitment to work towards being successful.

(**Update**): I wasn't planning on writing more in my chapter, but I decided to expand my knowledge in the lottery by adding more games to study. I wanted to see if I had the same experience with other state and national lotteries. Like I mentioned before, I assumed that the lottery games would be like the ones I've been doing for years, but the only way to find out is to put in the work and see, so I did. Yes, it was A LOT of work adding more to the games I already do. But the research was worth it. My hypothesis was correct. Every game I have extensively studied has been the same as the things I've explained to you.

Whether its combinations fitting together based on how many times numbers have come out. To the numbers being close with how many times they come out yearly. I also visited lottery facilities to observe the process and talk with lottery officials. It's a very thorough process. They truly believe the numbers selected are random. I'm still skeptical of that being true. There has to be some method to the system that makes the numbers come out the way they do. I have a theory about the selection process, but until I have facts, I don't want to talk about it publicly until I know for sure. Do I think most lotteries are rigged? No. I've seen enough evidence where I believe it's legitimate. Do I think the process can be rigged? Yes! I feel that way about a lot of things, whether it's stocks, elections, lotteries, etc.... All types of things can manipulate a process. So, it's hard to 100% trust anything. A lot of people like to talk about the lottery being rigged, especially when they lose. But I'm giving you my honest opinion, I think they try to keep it as fair as possible. If you do it the way I've taught you, the results will show that you can win.

So, like I said before, the best way in my opinion to truly learn the lottery is to go through the past results and add everything up. I'll say it again, yes, it's a lot of work. I like to go back the previous year or two, so I have a database of information. That way you can visualize the numbers and learn about each one. You'll see numbers that go together often. You learn different trends and patterns. Certain games have numbers that go together in a row often, for example, 1-2-3, or 35-36-37. Once you understand the trends and patterns of the specific game you're playing, that can help you figure out what numbers to go with. You learn by studying.

When you've seen the numbers so many times, you memorize a lot of key information. Don't be lazy and just go to a website that tells you how many times a number came out, and which ones have come out the most and least. Telling you this number is "hot", or this number is "due." Numbers come out when it makes sense for it to come out. You're cheating yourself out of learning if you only want to take shortcuts. If you really want to understand the lottery and have the potential to win consistently, know how to put the numbers together. It's easier than you think.

One of the reasons I like to tally all the numbers is it gives a good visual of all the numbers together. So, if #28 and #29 have both come out 13 times on a Thursday, I see it. Or if #7 has come out 11 times and #8 has come out 12 times. It gives me a visual that it might be time for #7 to come out to be even with #8. That happens with the numbers often with all the games I've studied. That's why I say the numbers stay pretty even with how many times they come out.

I'll give you some more examples of other random games I decided to study. Once again, I'm not going to tell you the numbers that came out, just how many times so you see how the combinations fit together. So, this one is <u>Cash 4 Life</u>: It's national lottery in 10 states. Here is a full week example so you can see what it looks like Sunday-Saturday: **(Sunday): (33), (29), (26), (26), (23), CB- (94). (Monday): (18), (33), (33), (28), (28), CB- (76). (Tuesday): (41), (24), (37), (28), (27), CB- (95). (Wednesday): (36), (37), (33), (29), (35), CB- (94). (Thursday): (34), (20), (28), (29), (30), CB- (95). (Friday): (37), (24), (27), (34), (29), CB- (96). (Saturday): (42), (28), (30), (31), (28), CB- (96).**

This one is <u>Lotto America</u>: It's a national lottery in 13 states, that comes out Monday, Wednesday, and Saturday. Here is a weekly example. **(Monday): (15), (18), (15), (19), (11), (SB)- (11). (Wednesday): (17), (14), (13), (17), (14), (SB)- 19. (Saturday): (14), (14), (18), (15), (16), (SB)- (10).**

This one is <u>Texas Two Step</u>: It's a lottery game in Texas that comes out Monday and Thursday. **(Monday): (25), (29), (30), (28), BB- (8). (Thursday): (16), (28), (18), (25), BB- (7).**

This one is <u>Fantasy 5</u>: It's a lottery game in Arizona that comes out Sunday-Saturday. It's just like Rolling Cash 5, my favorite game. On this example you can see two different ways why the combination works. Total numbers added up from previous years: **(82), (83), (81), (101), (83).** Or total numbers for one specific year: **(41), (42), (43), (49), (48).**

This one is <u>Pick 2</u>: It's a lottery game in Ontario Canada, that comes out Sunday-Saturday. I wanted to see if the lottery is similar in another country. Once again, many numbers are close with how many times they came out. On this midday example; **Monday (79), (79), Tuesday (78), (77).** Pick 2 can be easy to predict if you're following the trends and patterns closely every day. There are numbers that come out together very often. I wish my state had Pick 2; I could easily make an extra $20,000. If you have Pick 2 where you live, I would suggest starting with that, if you're trying to learn the lottery. The numbers are 0-9. But you only need to figure out two numbers that go together. Once you've mastered it, you can take the training wheels off and focus on other games with more numbers and larger payouts.

This one is <u>EuroMillions</u>: It's in nine countries – Austria, Belgium, France, Ireland, Luxembourg, Portugal, Spain, Switzerland, and the United Kingdom. It's played Tuesday and Friday. Once again, I wanted to see if their lottery is like the United States. The answer is yes, based on everything I kept track of many numbers are close with how many times they came out. I was able to see how the combinations fit together. Here is an example of two straight weeks. **(Tuesday): (11), (9), (12), (6), (10), LS- (18), (12). (Friday): (9), (13), (12), (8), (12), LS- (18), (17). (Tuesday): (12), (12), (11), (10), (11), LS- (21), (14). (Friday): (11), (10), (20), (15), (8), LS- (12), (13).**

These are just small examples to show you why combinations go together. I could give you hundreds of examples just like this for all the games I keep track of. I've spent over 10,000 hours studying lottery games. I have various ways where I'm able to put numbers together. I immediately know what to look for. At this point I only need about 30 minutes to figure out what numbers I'll play. I used to spend at least two hours trying to figure out what to go with.

My goal in 2024 is to continue to study more lottery games in other areas and see if I can become an expert in those as well. I'm in double digits now, of games that I can confidently say I understand and be able to show my work to explain why the numbers came out the way it did. That doesn't mean I would win every time. I'm not perfect, nobody is. I at least have enough information I can use to help get some numbers. I enjoy the process of the lottery, more than I enjoy winning money. I love seeing why the numbers fit together. So many people think they know about the lottery. There are a lot of so-called experts who lack substance with explaining how to win.

There are mathematicians who haven't put in significant hours to be able to talk about the lottery. Then you have random naysayers that repeat the same talking points that the lottery is rigged, it's a stupid people tax, you can't predict the numbers, and all the other negative comments. I'll say it again. I encourage everyone to truly follow the lottery, taking notes and studying every day. Just pick one game. I'm not doing anything, that other people can't do. The difference is, I'm one of the few people dedicated to learning the games and trying to understand why the numbers come out the way they do. I would've never gotten to this point had I had the same ignorance in early 2016. Don't judge the lottery if you don't actually know what you're talking about.

I wasn't going to include this either. But I'll show you how I structure the Pick 3. Just in case I wasn't clear enough earlier when talking about it.

I have separate pages for midday and evening. I tally up every time a number comes out. I'll use basic numbers as examples.

(August 2023)

0- |||

1- |||

2- |

3- |

4-

5-

6-

7- |

8-

9-

On the other side of the page, I write the date and numbers that came out.

8/1- 123 (Tues)

8/2- 101 (Weds)

8/3- 7<u>00</u> (Thurs)

Obviously, I'm going until 8/31.

On the next page I have a specific page for tally marks for each month. A lot of times there is a repeat number the next day. If you can identify what the repeat number will be, then you can try to work from that information in your notes. Let's say you're trying to figure out what the numbers will be for Thursday. I'm looking at the numbers that came out Wednesday. Based on information in my notes. Let's say over the past year the #1 has come out 100 times. Then I'm looking at numbers that could fit with it. I see that the #0 has come out 98 times over the past year. I'm making that one of the options as a potential match.

After looking over all my notes, I decided that 00 makes the most sense as two of the numbers. So, #0 will catch up with #1 and both will be at 100 times. They'll also be even for this specific week with how many times it came out. Now I need to figure out what that third number will be. For pick 3 I have 8 years of notes, so I have plenty to look through. Let's say I'm looking at how many times numbers have come out over the past 8 years. On that page if #7 has come out 750 times and #0 also has 750. That could be a potential match. So, that's how I piece together playing 700. If it happens to come out, now #7 is at 751 and #0 is 752. I'm just giving you a basic example, so you'll see what it's like. Numbers don't always fit perfectly together based on how many times they have come out. Sometimes you'll have to look at other potential information, I'll show you further down.

There are many times, I'm looking through my notes and I honestly have no idea what to go with, nothing stands out. If I'm not confident I have something, I don't play the game that day. I never feel the need to play just to play. Just collect the data and wait until you have something figured out that makes sense. Sometimes you'll win. Sometimes you'll get a couple of numbers. Sometimes what you thought will be completely wrong, it happens. Just keep learning every time.

(August 2023)

(8/1- 8/5)

0- |||

1- |||

2- |

3- |

4-

5-

6-

7- |

8-

9-

(8/6 - 8/12)

0

1-

2-

3-

4-

5-

6-

7-

8-

9-

I keep going each week for the month. Sometimes during the week some numbers won't come out. A lot of times most numbers will come out 2 or 3 times during the week. But it varies, sometimes one of them might be 6 and another will be 0, and the rest in between. I only use this page to get an idea of how many times numbers come out each week. This information doesn't really impact how I pick numbers; it's the least important page.

On the next page I have the 2023 1st number. So, I only tally each first number midday, and I have another one for the evening.

0-

1- ||

2-

3-

4-

5-

6-

7- |

8-

9-

Next page is the 2023 2nd number.

0- ||

1-

2- |

3-

4-

5-

6-

7-

8-

9-

Next page is the 2023 3rd number.

0- |

1- |

2-

3- |

4-

5-

6-

7-

8-

9-

I use these pages to help me see when numbers might be due. It also helps me decide when it might be a good time to play numbers straight, or something might stand out where it could be time for triple digits to come out. Throughout the year a lot of things will stand out with information you can use. I don't play pick 5 anymore. But when I did, on my pick 5 version this was a good way to figure out what numbers to go with. Whether it's: First four numbers in exact order. Last four numbers in exact order. First three numbers in exact order. Last three numbers in exact order. First two numbers in exact order. Last two numbers in exact order.

Next page is the 2023 total number. This is the most important page. At the end of the year, I will continue to use these totals for the next year. I'm just using random numbers to give an example. This is what I mean when I say numbers stay pretty close together with how many times they come out.

0- 45

1- 49

2- 57

3- 52

4- 39

5- 41

6- 56

7- 44

8- 60

9- 47

The next pages I have are the numbers that came out throughout the years. So, for example (Past Year), (Past Two Years), (Past Three Years), etc.... This is what mainly helps me figure out what combinations to go with. Like I've said before, in all the games this is usually why the combinations fit together like they do.

The next pages I have each day the games are played and again individually for the 1st, 2nd, 3rd numbers and total times that day. I do this for midday and evening.

(Tuesday 1st #)

0-

1- |

2-

3-

4-

5-

6-

7-

8-

9-

(Tuesday 2nd #)

0-

1-

2- |

3-

4-

5-

6-

7-

8-

9-

(Tuesday 3rd #)

0-

1-

2-

3- |

4-

5-

6-

7-

8-

9-

(Tuesday 2023 Total)

0-

1- |

2- |

3- |

4-

5-

6-

7-

8-

9-

So, that's my process for Pick 3 midday and evening. It can be a lot to keep up with, but it's worth it to me because it works. I would never put this type of information out there if it didn't work. My goal is to help people, not mislead or scam people selling my 'system.' Like I said, the only system is the lottery system itself. All I'm telling you to do is use the information that's already available to figure out what combinations to put together.

For years, I've heard people talk about using a wheeling system, I never bothered to look at it until I decided to write this book, just to see what the hype was about. Keeping it real, I would NEVER play the lottery that way. If it works for people, I'm not going to knock it, keep doing it. I like data and information. I'm detail-oriented and like an organized way to do things. Wheeling doesn't help me truly learn as much I can, to win consistently.

My way won't be for everyone. This is for the serious dedicated person. If you're not focused, organized, patient, detail-oriented, or you lack commitment to keep track of information. Don't bother doing it, you're just wasting your time. No matter which game you're playing, if the information you're keeping track of isn't accurate, you'll be wasting your time, because the combinations won't work like it should. I've given you the blueprint, what you do with it is on you or if you're trying to do it with other people.

Honestly, I don't feel like showing you each specific process for every game I follow. You should be able to get the concept based on everything I've explained throughout the chapter. It wasn't until June 2023, where I really started wondering what people say about the lottery. I'm usually focused on what I have going on, so I don't pay attention to what others say or how they do the lottery. There are people on websites, message boards, and video channels giving out information about the lottery and don't study it or play it. They'll say things that are clearly wrong or misinforming people. I didn't even write down each one that I could prove wrong. I'll just mention what I remember.

The lottery is based on randomness: Debatable, but I lean more towards there's a method to the system. Whether they use a pseudorandom number generator or other factors. Not every lottery entity is the same, and there are many different games worldwide. I'm still trying to confirm something I noticed during the lottery selection process at the places I observed. That will be in a future book if I do one. I want my analysis to be based on facts, not speculation.

Lotteries are just like flipping coins: I've heard many people make a comparison between lotteries and flipping coins. I believe the lottery is more like putting together a puzzle. There are many pieces involved when putting the combinations together. Most games have 9 to 95 numbers to choose from. I haven't studied every lottery game in the world. But from extensively studying different games worldwide, the evidence has shown the numbers come out for a logical reason. It's not as random as people think.

Past results have no impact on future results: I've made it clear that's not true. I have 8 years of data that shows that's the main reason the combinations go together. It doesn't matter whether someone believes me or not. I can at least back up my points consistently with facts. Professor so-and-so can't. Media outlets will interview a professor that admits they don't play the lottery or follow it closely. Yet, they have the answers on how the lottery works and how people should play. Watch the confidence shrink if you ask someone if they're willing to pay a million dollars if they can't consistently prove their theory showing work.

Lotteries are designed to prevent patterns and trends: I don't know whether it's designed that way or not. That might be the narrative put out there. But again, I've shown that's another key thing that happens often with many of the games I've studied. If you're truly studying each individual game, most of you should be able to notice patterns and trends. Some of them repeat the same specific things each month, year after year. Just like the lack of randomness, it might be by design.

Currently, I'm trying to learn about some of the companies that handle gaming systems around the world like Scientific Games, International Game Technology (IGT), Gaming Labs International, and Smartplay International. We're told there are algorithms designed to produce numbers that lack any predictable pattern. We're told millions of dollars are spent on updating systems. Both things may be true, however it seems to me if the patterns and trends haven't changed since I've been following the lottery it must be by design. Theoretically, I shouldn't be able to predict that certain things will happen.

When lottery systems do test, I think they're more focused on making sure the same numbers don't pop up frequently in the draws as opposed to other factors like combinations that go together and repeated trends during specific times.

One of the main talking points is that the lottery is unpredictable, if people believe it and repeat it, they probably save millions by not having to constantly change the system. That's just my opinion. I'll continue to research and investigate the behind-the-scenes process.

The lottery history is there, you can study any game yourself. I encourage you to do it. Don't just glance, truly examine everything. Take notes on all the repeated trends and patterns. Days, dates, months, years, all of it. This is applied to games run on systems and not selections from a ball machine. That might be less predictable. But you should study every lottery game anyway, that's how you learn.

The more tickets you buy, the more chances you have to win: I won't debate this point. As you'll read in other chapters, this has worked for some people. What I will say is, it's not a strategy I would do. Other than when I first started playing the lottery. I've been a very disciplined player. If I'm not confident (meaning almost positive) that I have enough solid numbers to win money, I won't play. I focus on getting numbers right, not the jackpot. Instead of throwing a bunch of money trying to win the $1,000 a day jackpot prize. You could still do that with winning $500 with Pick 3 midday, and $500 with Pick 3 evening. Be smart and strategic with how you play. I just gave you the blueprint earlier.

Only do lottery for entertainment purposes: I'm not going to tell someone how they should approach it. If someone wants to use the lottery to improve their life, just be smart about it. I've tried to stress this point throughout the chapter. If you're smart with how you play, it can work out for you whether it's for entertainment or changing your circumstances. I can't guarantee you'll win. I can only help you with what has worked for me. In my opinion, stocks aren't necessarily better than the lottery. Whatever works for you.

Spread numbers out mixing odds and evens: Just like with "hot," "cold," and "overdue" numbers. Whatever game you're playing numbers will come out when it makes the most sense for the combinations. I've seen it over and over. Focus on putting the combinations together that makes sense based on the data you have. If it ends up being mixed, all even, or all odds, that's just the way it worked out that time.

One of the reasons you want to be careful when trying to pick numbers you think are due is that you might wait all year for it to happen. A perfect example is that there was a Mega Ball that was the only one that didn't come out until the last draw of the year. It happened because it made sense for it to come out to finally be even with the Mega Ball number before it, with how many times they came out over the years.

So, if you kept thinking it's due every time you played, you literally would've waited all year. Sometimes numbers don't come out all year. Or some numbers won't come out in a specific order on a certain day. For example, with Pick 3, #5 might not be the first number on a Saturday all year long. That's why I like my individual day/number order pages, so I can pick up on things like that. It's valuable information.

Try picking sets of numbers that reduce matching other people's picks, so you can avoid splitting the jackpot: Too many people worry about what other people might pick and what people say about the probabilities and chances to win. It doesn't matter, I'll say it again. Focus on getting numbers right. Whatever makes sense for the combination, if #2 is one of the picks, who cares that 100 million other people might pick #2. As for splitting the jackpot, Lewis and I were laughing about this, people complain about potentially having to split $500 million dollars when they spent $20 on auto-picks. You just made an insane profit without doing work and still making a fuse about taxes and splits. C'mon seriously?

People that sell books about the lottery and gambling are scammers: I can't speak for other people's motives. I'm sure there are people doing it to scam people. I'm sure there are others who truly believe that they have a 'system' or figured out the lottery. I've been passionate about the lottery since 2016, I've learned a lot over the years, and I want to share that knowledge. I know what it's like to live paycheck-to-paycheck. There are times when people need assistance and have to set up a GoFundMe page. I believe people can utilize the lottery if they do it right.

Some people will look at the title of this book and think it's about making millions of dollars gambling. It's more about personal goals and how you define success. Success to you could be making an extra $500 a week or month. One of the chapters in this book might help you get there. I wouldn't ruin my credibility over $10-$20 especially when I have no idea whether anyone will actually buy this book or like it. The truth is, I don't care whether people buy my books or like them. I publish the things that I'm interested in. My life doesn't change whether someone buys my book or not. Be careful with anything you purchase. I don't want to see anyone get taken advantage of.

You're probably wondering how many times I've won. Or how much money I have won over the years. It doesn't matter. My success has nothing to do with how well someone else might do. I don't want to be defined by a number, like "seven-time lottery winner" or 'that's the guy that won $20 million dollars.' Plus, it's personal information. If most people could stay anonymous with lottery earnings, they would probably choose that option. The fact that I've been willing to share this much information is more important. I could've easily kept quiet and kept the knowledge to myself. I would rather my lottery legacy be known for helping others make life-changing money.

I will say this, I've won thousands of times, and I've lost thousands of times. I compare it to the most successful MLB managers. I'll use John McGraw for example. He's ranked third all time, he won 2,763 games, lost 1,948 games and tied 58 games. He has a 58.6-win percentage. A win for me is making more than I spent. So, if I spent $1 and won $10, it counts as a win. If I break even, it's like a tie. I'm good at the lottery, but I'm far from perfect.

2024 will probably be my last year doing the lottery as much as I do. I want to focus on other things. I want other people to continue what I've started and keep it going as long as possible. I expect they'll change the lottery system eventually if millions of people are successful doing it my way. If the lottery system has to change because so many people are making money, great! Turn that "stupid people tax" into the smartest decision you ever made.

Honestly, I don't expect most people will be as dedicated as I've been. Some of you might take bits and pieces of what I've done and that's fine. Realistically, if you spent $2 on a Mega Millions ticket weekly. $2 on a Powerball ticket weekly. 50 cents on Pick 3 midday or evening for a total of $3.50 weekly. That's $7.50 weekly, $30 a month, and $360 a year. If you get to the point where you can consistently win at least $10 on a Mega Millions ticket weekly, $7 with Powerball weekly, you win $41.50 once a week with Pick 3 midday, $83 once a week with Pick 3 evening. That's $141.50 a week, $566 a month, and $6,792 a year.

Of course, it's not a $670 million dollar jackpot win. You have the potential to get that, but I'm just giving you a modest example that if you consistently win with small payouts, that can still help towards a college fund, bills, a nice vacation, use it for other investments, whatever you decide to do with it. You can play smart and make money, the lottery doesn't just have to be about a large jackpot.

For me personally, I like to give myself goals. Sometimes it might be $1,500 a day. $300 from Rolling Cash 5, $200 from Lucky for Life, $500 from Mega Millions, $250 from Pick 3, and $250 from Pick 5. That's five different games for $6 total, to win $1,500. All it takes is $300 a day to make $100,000 a year. That's what I try to get people to understand. If you're smart and strategic with how you play, you can make a nice living.

I've had over 600 opportunities to play Mega Millions since I've been keeping track of it over the years. I've played it less than 10 times. I just keep collecting data and play when I'm interested. I prefer to focus on winning daily games with fewer numbers; 95 is a lot to piece together for two days. It wasn't until the middle of 2023 that I finally started to focus on studying Mega Millions and Powerball to really learn them and know them like I do with other games.

When I do play them, I've won a decent amount. So, my point is you don't have to play all the time if you're not interested in certain games or just continue to study until you're ready to play. Like I've said before, you should probably focus on one game first and master it, then add more if you feel like it. Don't overwhelm yourself. You could also just focus on one specific day and master it. If you just take notes on Friday's Mega Millions, continue to study it and learn the trends and patterns and use that information to help you win money.

If you get to the point where you're extremely committed to this for years. This is what it looks like; my thick folder of lottery notes for all the games I study. Yes, I end up going through all these throughout the year. That's why I know what I'm talking about. I have it backed up on the computer. But keeping it up to date on paper is better for me. It's all organized by game, so I get through it fast. The only time it's a hassle is New Year's Eve when it takes 12 hours to prepare for the New Year and start off fresh.

(**Author's Note**): The other examples are from the first edition of the book in 2023. I'm including this in the updated 2025 edition. I used this in a lottery class I taught in Florida. I've shown the examples of why the combinations went together. Now I'll show a monthly example of the numbers that came out. The underlines are numbers that have gone together multiple times throughout the month.

The * is to show that 12 times a row was completely filled. Underlining has helped me win a lot of money. When you can narrow down what numbers will go together you should be able to get at least two or more numbers. I also wanted to show you about the blocks of numbers being used and adding up numbers like I mentioned before.

With this game you don't necessarily need to know how many times a number came out to put the combination together, I would still do it to have the information. But just seeing how easy they make it with the blocks of numbers you could make a profit by just fitting some of the numbers together when you see the trend. I studied some of the Florida games, it's one of the easiest states to win money in my opinion.

Like I told the class, some lottery games are begging you to win money by how easy they make it sometimes, you just need to know what to look for. That's why I tell people odds and probabilities don't matter if you're willing to learn. You can separate yourself from the people that would rather repeat talking points that you can't figure out the lottery.

(Fantasy 5 Florida Midday January 2025)

Numbers for the draw (1-36)
    (1/1): 4,5,11,19,35 (74) (Weds)
    (1/2): 3,4,10,16,20 (53) (Thurs)
    (1/3): 2,10,16,21,26 (75) (Fri) *

(1/4): <u>10</u>,<u>13</u>,<u>22</u>,<u>23</u>,28 (96) (Sat) *

(1/1-1/4): 2,3,4,5,10,11,13,16,19,20,21,22,23,26,28,35

(1/5): <u>1</u>, <u>6</u>,15,<u>21</u>,22 (65) (Sun)

(1/6): <u>15</u>,20,<u>24</u>,<u>26</u>,28 (113) (Mon)

(1/7): <u>7</u>,13,<u>23</u>,30,35 (108) (Tues)

(1/8): <u>4</u>,9,<u>12</u>,<u>13</u>,<u>21</u> (59) (Weds)

(1/9): <u>10</u>,<u>19</u>,<u>20</u>,25,28 (102) (Thurs)

(1/10): <u>1</u>,<u>8</u>,11,<u>21</u>,23 (64) (Fri)

(1/11): <u>8</u>,13,14,<u>21</u>,27 (83) (Sat)

(1/5-1/11):

1,4,6,7,8,9,10,11,12,13,14,15,19,20,21,22,23,24,26,27,28,30,36

(1/12): <u>6</u>,<u>12</u>,<u>29</u>,32,36 (115) (Sun)

(1/13): <u>13</u>,16,<u>17</u>,<u>27</u>,<u>31</u> (104) (Mon)

(1/14): <u>1</u>,<u>6</u>,<u>26</u>,<u>29</u>,<u>34</u> (96) (Tues) *

(1/15): <u>22</u>,<u>24</u>,<u>26</u>,<u>28</u>,<u>29</u> (129) (Weds) *

(1/16): <u>2</u>,<u>15</u>,<u>18</u>,<u>27</u>,28 (90) (Thurs) *

(1/17): <u>4</u>,6,<u>19</u>,<u>28</u>,<u>29</u> (86) (Fri)

(1/18): <u>2</u>,<u>4</u>,<u>7</u>,14,<u>20</u> (47) (Sat)

(1/12-1/18):

1,2,4,6,7,12,13,14,15,16,17,18,19,20,22,24,26,27,28,29,31,32,34

(1/19): <u>2</u>,<u>4</u>,<u>17</u>,<u>21</u>,<u>26</u> (70) (Sun) *

(1/20): <u>4</u>,<u>7</u>,<u>26</u>,<u>27</u>,<u>33</u> (97) (Mon) *

(1/21): <u>6</u>,<u>12</u>,<u>18</u>,<u>27</u>,<u>31</u> (94) (Tues) *

(1/22): 17,<u>19</u>,<u>20</u>,30,32 (118) (Weds)

(1/23): <u>4</u>,<u>10</u>,<u>19</u>,<u>20</u>,<u>33</u> (86) (Thurs) *

(1/24): <u>5</u>,<u>7</u>,<u>10</u>,<u>23</u>,<u>32</u> (77) (Fri) *

(1/25): <u>13</u>,<u>23</u>,<u>29</u>,<u>32</u>,<u>34</u> (131) (Sat) *

(1/19-1/25):

2,4,5,6,7,10,12,13,17,18,19,20,21,23,26,27,29,30,31,32,33,34

(1/26): <u>2</u>,9,<u>26</u>,<u>27</u>,<u>29</u> (93) (Sun)

(1/27): <u>2</u>,<u>10</u>,<u>13</u>,<u>24</u>,<u>29</u> (78) (Mon) *

(1/28): <u>2</u>,<u>13</u>,<u>17</u>,18,<u>20</u> (70) (Tues)

(1/29): <u>12</u>,<u>13</u>,15,<u>19</u>,<u>33</u> (92) (Weds)
(1/30): <u>5</u>,9,<u>10</u>,15,36 (75) (Thurs)
(1/31): <u>12</u>,<u>13</u>,14,<u>22</u>,<u>29</u> (90) (Fri)
(1/26-1/31):
2,5,9,10,12,13,14,15,17,18,19,20,22,24,26,27,29,33,36

When talking about Pick 5, I also gave an example of an individual day in Ohio. So, I'll also share this with you as well to get a visual. This is how many times numbers came out.

(Pick 5 2024 Sunday Evening 1st #)
(0)- 4
(1)- 5
(2)- 5
(3)- 4
(4)- 9
(5)- 5
(6)- 8
(7)- 7
(8)- 4
(9)- 6
(Pick 5 2024 Sunday Evening 2nd #)
(0)- 7
(1)- 5
(2)- 5
(3)- 7
(4)- 3
(5)- 10
(6)- 4
(7)- 3
(8)- 2
(9)- 6

(Pick 5 2024 Sunday Evening 3rd #)

(0)- 3

(1)- 3

(2)- 8

(3)- 5

(4)- 2

(5)- 9

(6)- 7

(7)- 4

(8)- 7

(9)- 4

(Pick 5 2024 Sunday Evening 4th #)

(0)- 2

(1)- 3

(2)- 4

(3)- 9

(4)- 1

(5)- 6

(6)- 6

(7)- 8

(8)- 4

(9)- 9

(Pick 5 2024 Sunday Evening 5th #)

(0)- 6

(1)- 5

(2)- 3

(3)- 6

(4)- 6

(5)- 3

(6)- 5

(7)- 5

(8)- 4

(9)- 9

(Pick 5 2024 Sunday Evening Total)

(0)- 22

(1)- 21

(2)- 25

(3)- 32

(4)- 21

(5)- 32

(6)- 25

(7)- 27

(8)- 21

(9)- 34

Hopefully, I helped you understand the lottery better. Feel free to email me with comments and questions. Sometimes I'm willing to consult. I'll try to reply when I can, LanceCares@gmail.com

K.J.

Since I was 5 years old, I've always been a huge sports fan. I was a great athlete as well. In high school, I was a three-sport athlete. I had 22 scholarship offers to play in college. My senior year in school, my friend said we could make some easy money by robbing a pizza man. His older brother and his friend got $200 cash, $800 in jewelry, and a $150 pair of shoes from robbing a pizza man at gunpoint. I needed some money ASAP, so I wanted to do it. We had it delivered to an apartment complex at a random address. We wore glasses, hats, and hoodies, doing as much as we could to hide our faces. We're waiting by the stairs of the address, not the actual apartment. The pizza person comes up, and my boy hits him with the baseball bat. We demand he give us everything he has, including keys to the car.

We grabbed everything we could and took off in it. We drove it a couple blocks away. We hopped out on foot and ran as fast as we could to a stolen car we had stashed away in an alley. While he's driving, I'm looking to see what we had; it was less than $50; it was a bust. We did all that for nothing. That only motivated us to do it again until we got it right. We broke into some houses and stole a bunch of stuff. We had a lot of valuable stuff we were able to sell off. A couple of things we tried to get rid of ended up being sold to an undercover cop. So, we were charged for that. My boy dropped his ID near the stolen car when we robbed the pizza guy. We were linked to that crime. I was sentenced to four years in prison. My athletic scholarship opportunities were gone.

While I was in prison, I learned how to gamble. One of the older guys in there used to run an underground casino operation in the 80s. He taught me everything he knew about gambling. Whether it was card counting, team playing at table games, or understanding sports lines, I made sure I understood everything he was telling me. Because when I got out, I intended to use it. He didn't have kids, and I was the only one in there who listened to him and had conversations as much as we could. He loved telling stories of his glory days. Other people weren't trying to hear it. He appreciated that I took an interest. Since he was going to spend the rest of his life in prison with no possibility of release, he wanted me to have his hidden cash when I got out. He had $40,000 buried in the woods near one of his old homes. If I could find it, I could have it. He explained it to me many times, so I would remember where to look.

When I got out, the first thing I did was go to the city and location, looking for it. There were so many damn trees, weeds, and branches that I thought I might spend the rest of my life trying to dig up this money. I looked in every spot possible; nothing looked like how he described it since it had been so long. Finally, on the 9th day, after digging deep in so many spots, I found the buried bags. He wrapped them so nicely that none of the money was wet, dirty, or anything. I kept grabbing bags and taking them to the car I borrowed from my cousin. I got the nicest hotel I could find and stayed there for a few days, working on a game plan for what's next. I wanted to flip this $40,000 into $400,000 by gambling with the knowledge that I was taught. In his heyday, he was making $10,000 a day. I wanted to do the same. Some people spend four years in college to get a degree in something. I got my degree in advantage gambling and was paid for it. One of the first things he taught me was that the casino is not looking for the customer's best interests; they want to take advantage of you as much as possible and are not looking to play fair. That's why they give you perks, trying to butter you up and let your guard down. Next thing you know, you lost $20,000. So, you need to go on the offensive as an advantage player and take advantage of them. They want to maximize their profit, just like you. Get them before they get you. He made a lot of money getting people before they got him. Whether he was the house or the player.

I started playing blackjack at hole-in-the wall casinos with lower limits, just trying to get my feet wet and see if I learned what he taught me. I was easily beating the house, winning hundreds and low four figures; every technique worked. Especially the Ace Sequencing. Trying to explain all of it in a way that people will understand is hard. You have to visualize and see it point-by-point to know what I'm talking about and where it would make sense. A lot of people still don't understand when you show them while explaining. It's very technical stuff. I remember when he first taught me, it felt like I was back in school trying to learn trigonometry or some other complicated subject. Sometimes it takes a long time to go over it constantly until you finally memorize it and process it. You learn as you go; experience is the best teacher as an advantage player.

It was time for me to raise the stakes, go to bigger casinos, and play for more money. There's a lot more pressure and sometimes it's harder to focus in high-stakes places. There's security watching you like a hawk, people walking the floor, and cameras everywhere. Pit bosses and hosts want to chat with you and be a distraction. The dealers are a lot faster and more precise. At the smaller, less impressive casinos, the dealers had a tendency to flash some cards, giving me the advantage, I needed. I like when women flash me, but I love when the dealer flashes me. I took some hits at first while playing high-stakes blackjack. I traveled to different casinos, experiencing different dealers and atmospheres.

I learned it's easier for me to win when there's a female dealer. In some casinos, I would win low five figures. I made recap lists, seeing what worked when I won big and what went wrong when I lost often. I took notes on which casinos fuck with me the most. Asking for my ID, asking me too many questions, and trying to kick me off the table when I'm doing well. I wrote down the shifts of my favorite dealers and the dealers I hated, trying to figure out the best routines for when to play. I wrote down how often some places swap out. I wrote down the best casinos to play at and the worst. The worst casinos to play at, in my opinion, are Indian casinos. I'll never go to those again. The best for my personal success was Reno, Nevada, with winning money and raising my bets without issues. Also based on the checklist of notes, things like not really messing with me, going against the dealers, taking care of me with perks—that type of stuff.

It's a roller coaster to be a high-stakes gambler. There are times I win over $50,000 and then lose over $60,000. Casinos hate it when you cash out big. Sometimes they look for every way not to pay out the money they owe you. Or if you're hitting a winning streak, that's when they try to kick you off the table, giving any bullshit reason why they want you to leave. I've had casinos steal my chips worth thousands of dollars. I try to go to 10 to 15 different casinos a month and keep rotating around each region in the U.S. I like to go a while before returning to some casinos frequently. Anything you can do to avoid putting them on your radar for unwanted attention is better for you. Especially if you're a known card counter or casino hustler.

The best day I had was mid-six-figures. On average, when I'm doing well, I make about $5,500 a week. When I'm doing badly on average, it's five-figure losses. Professional gambling at casinos involves a lot of hours of playing; it's stressful, and you'll most likely lose more money than you win. If you're fortunate enough to win big, the smart thing to do is walk away on a high note. But most of us don't do that. We always want more; that's what drives us to do this for a ridiculous number of hours and take big risks with big bets. Blackjack is the main game I do, but I do other things as well. Roulette, craps, and poker. I've also become a big sports bettor. I wager on a lot of games. I've made millions of dollars from gambling, and I've lost millions of dollars in my 14 years as a professional gambler. Currently, I've made more than I lost. The grind of gambling starts to wear you down. I'm not as enthusiastic as I used to be. On my 15-year anniversary, that will be my last time doing it. I'm putting in all the chips and retiring after that, win or lose. It's time to move on to other things.

I have a wife and kids now. I have a farm. I have stocks and investments. I've made enough that I'm financially stable. I was given a gift of knowledge and $40,000, and I turned it into $7 million with gambling, stocks, real estate, and investments. My mentor passed away; I'll never be able to repay him for the opportunity he gave me to change my life from a criminal to a successful gambler and entrepreneur. I've been one of the fortunate ones, but I'm not going to lie; it was a long and challenging road. I almost went bankrupt several times. This is absolutely a high-risk, high-reward industry. I spent half of it in the red. Be prepared to fail a lot at gambling. You'll probably be harassed a lot by casinos and family members if you win. Keep your profile as low as possible.

Lewis

When the lottery got large, people at my job would put their money together, and we would buy as many tickets as possible. One day, on a lunch break, I was talking with one of my co-workers. I was telling him that instead of waiting until the lottery got large, we should just do it every day and see what happens. Let's crowdfund the lottery, like people do for business, donations, and other things. The lottery is already cheap, but if we're all going in it together, the risk is a lot less, and we have an opportunity to make extra money if we win. I thought about this before I heard of the movie *Jerry & Marge Go Large* or heard of them. It seems like common sense to me that the more tickets you buy, the more opportunities you have to win. So, we got a group of 4 people to go in together for all the tickets we bought, and we would split it evenly if we won. If the jackpot is a million dollars and we divide that among four people, that's $250,000 pre-tax for a $10 investment on my part. One of the dumbest things I hear people say, is they wouldn't want to share the winnings. I would never complain about getting $100,000 because I had to split the winnings, and I only spent $10 to get it.

To make sure there weren't any disputes in the future, we all signed a contract with agreed-upon terms. Whoever bought the tickets had to take pictures and send them to the group to make sure they were purchased. When the tickets were purchased, we had a safe with a camera on it to see who opened it. Every time it was opened, we got notifications on our phone. We trust each other, but you can never trust anyone 100%. Better to be safe than sorry. We bought Powerball tickets, Mega Millions tickets, and scratch-off tickets.

First, we lost more than we spent. We knew we weren't going to get rich overnight, so we didn't get discouraged and continued playing. About 4 months later, our luck started to change; we were finally making more than we spent. Our first big win was $100,000 on a scratch-off ticket. We increased our investment, bought more tickets, and even traveled to nearby states to play some of their lottery games. Within the first year, we won what we consider big wins seven times. The big wins for us were at least $5,000. We quit our jobs and formed our own LLC. We considered going our separate ways; we all had enough money to spend on our own. But we liked the partnership; we all invested in franchises together.

I went from making $48,600 a year at my job to becoming a multi-millionaire in two years. I live off lottery winnings, and all the money I make from business I save. I thought I had some original idea, then I started seeing promos about *Jerry & Marge Go Large* and reading more about what they did and their rivals. At least this proves that what we do works. I don't know how many other groups do this on a regular basis. We didn't spend our life savings; we started with less than $50 a month in investment. We didn't start doing more until we made way more than we spent before. We stayed with that strategy; discipline was a key factor. We didn't spend outrageous amounts of money; we didn't get discouraged when we lost constantly. When we did win big, we kept each other grounded and stayed focused on our strategy. One tip I'll give with scratch-off tickets, go to different stores in different cities in your state. That gives your group a better chance of purchasing a winning ticket. Go to your state lottery website and review how many remaining winning tickets are available on each game. When the large prizes get below 20 tickets left, try to buy as many as you can.

My advice to people is to make sure you find the right group of people to partner with. Make sure you have a contract that everyone agrees on. The lottery comes with a lot of losses if you're doing this daily. Be prepared to know you're not going to win in your first month. So, set a budget with the amount of money you're willing to lose. Always stick to a modest budget, win, or lose. 70% of lottery winners go bankrupt within a few years. Don't be stupid. Don't quit your job until you're in a financially secure position. In my opinion you're not a professional gambler until you're able to live off your winnings.

Stacy

On a Saturday night, I went to the carryout to get snacks before my husband, and I watched a movie at home. While I was at the register, I decided to buy some scratch-off tickets. I had never done it before, but some of the games looked interesting. I bought $20 worth of tickets. I remember the clerk saying, "Good luck!" I'm thinking to myself, 'I probably just wasted $20.' There's no way I'll be lucky enough to win my money back or make a profit from what I spent.

When I got home, I started scratching off the tickets and reading what to look for to see if I won. Some of them were a little confusing; I wasn't sure if I had won something or not. I would hand them to my husband so he could look at them just in case I missed something or didn't understand how the game worked. The first three tickets were losers. Then I scratched the fourth one and won $10. I was happy; at least I made half my money back so far. I continued scratching tickets.

On the seventh one, I won $30. I was getting excited; I just doubled my money. I start scratching the other tickets faster. I felt like a kid opening Christmas presents, just rushing to see what I got. On my eleventh ticket, I won $2. It was not as exciting, but hey, I still won some more money. I scratched a few more and didn't win. On my final ticket, I scratched it off; it was another one I didn't understand how the matching worked, and I thought I lost. My husband looked at it, reading the rules of the game. It turns out that when you get the bonus, you can double the amount. I had the bonus on my ticket, so I won $400. My first-time buying scratch-off tickets, I spent $20, and I won $442. I guess luck was really on my side.

My husband was trying to figure out if there was some type of strategy, we could use to know how many tickets we could buy that might be winners. We would look on the lottery website to see how many winning scratch-offs are left. So, if we bought a bunch of tickets in a row, at what point would one of them be a winner. We watched some YouTube videos of people who would buy 1,000 scratch-off tickets: we're like, 'Wow, that's excessive!' But we were interested in seeing if their strategy worked, especially with the people who bought all the same tickets.

In all the videos we watched, people won on some tickets, but nobody won more than they spent. That was a lesson for us to remember, so we don't go overboard on buying tickets. We decided to split buying 50 tickets and see how well we did with that first before trying more. We both bought the same number of tickets for each game. We wanted to know if there was an average number of tickets before there was a winning one. I scratched mine, out of my 25 tickets, I won fourteen times. $100, $40, $30 twice, $20, $10 twice, $5 twice, $2 three times, and $1 twice. So, I won $258. I spent $130; I was satisfied with the profit I made. My husband scratched his off; out of 25 tickets, he won six times. On the second ticket, he won $1,000. We jumped in excitement, hugging each other; we celebrated like we had won $10 million. He also won $20, $5, and $1 three times. We made over $1,000 in profit. It felt like we were dreaming; we couldn't believe our luck.

I started buying lottery tickets; maybe my luck will extend to winning $50 million with the Powerball or something. When I bought tickets, I broke even a couple of times, but mostly I lost money. I would still buy scratch-off tickets; most of the time, I won some money. We did the best on scratch-off games; we could double our money on the tickets. I enjoyed the scratch-off tickets, but my husband was determined to be a lottery winner. He would get jealous every time people won these large jackpots.

We watched a show about how lottery winners who won millions lost it all with dumb purchases and trusting the wrong people. He said that if we win, that will never be us. Watching those shows does make you want to play the lottery; it's exciting to think about what you'll do with the money if you win. I started playing the lottery again, and my strategy was simple. Just play the same numbers every time; I figured at some point they'd have to come out. I played Mega Millions and Powerball; it's $2 each. If I spent $4 to $8 the rest of my life just playing those, I felt like eventually the return would be worth it. I would play Mega Millions on Friday and Powerball on Saturday. I just played the same numbers for both. Most of the time I won $4, so I at least made my money back. One night, I was about to go to sleep. I checked what Mega Millions numbers came out. I checked my ticket and saw I had four numbers. I rubbed my eyes since I was tired and wanted to make sure I read it right. I really had 4 of the 6 numbers and won $10,000. I ran downstairs. My husband was watching a game, and I yelled, "I WON! I WON! I WON $10,000!" He also played and wanted to check his tickets; he bought several with auto-pick. He had a $200 winner, a $10 winner, and a $4 winner. We felt like it was only a matter of time before we won the jackpot.

We were traveling to go to a family reunion. While we stopped at a gas station, I decided to buy some scratch-off tickets. I scratched and was already seeing winning things on there; I knew this was a good ticket. When I finished, my eyes got wide. I just won $1 million dollars. I screamed with excitement. My husband looked over at me with a big smile on his face; he knew it had to be something big for me to have that much excitement. I couldn't believe it; I had him check the ticket to make sure. He said, 'Sorry, babe, you read it wrong.' I was so disappointed. Then he said, 'Just kidding.' (Laughing). I punched him on the arm for killing my vibe.

When you win a million dollars, it's not even close after taxes. But it's still a lot more than what I had. It was enough for me to quit my job. I didn't want to be like those lottery winners who go broke. So, I still kept a modest budget for spending. I increased my $4 to $8 a week for the lottery, allowing myself to spend up to $20 a week. With a mixture of lottery tickets and scratch-off tickets, I've been making more than I did when I had my full-time job. We have played lottery games a lot and frequently, especially the more we started winning. I'm not sure I would consider us professional gamblers. But we've had tremendous luck and success. I don't want to tell people to try the things we did because I don't know if you'll have the same success. Play for fun and hope for the best, but don't spend your life savings trying to win a million dollars. If you win, don't become the next person to blow all the money and go broke.

Dave

The way I've made my living most of my adult life has been in the sports memorabilia business. I started it with college roommates in 1999. We purchased all types of items—lots of great stuff. Memorabilia from Michael Jordan, Pete Rose, Wayne Gretzky, and so many others. We sold our business for millions of dollars. My cut was over $2 million. I started off playing fantasy sports for fun. Every league I was in, I would win. I wanted to turn this hobby into a profession. The more popular daily fantasy sports became, the more I wanted to make money from them full-time. I feel like it's an easy way to make money. The entry fees are low, but the potential to win is high.

I have millions of dollars in the bank, so I have plenty of time to devote 40 hours a week to fantasy sports. There have been times I've entered over 60 daily contests. My friends joke and say I know players better than coaches and GMs. My teams consistently have the best stats. I'm very good at identifying which players will do well. I know when to add and when to drop. I spend a lot of time researching and developing strategies. People love playing fantasy football; that's the most popular. I enjoy it as well. But the key is finding the best money-making opportunities with leagues and places to play. I learned a lucrative lane is with fantasy baseball, in addition to betting on baseball because there's so many games available. The first week, doing different bets and fantasy baseball, I made $18,200.

I went to a video game convention. That was a whole new world for me. It felt like Comic-Con for video games. I went to the NBA 2K room, and you had to pay $100 just to go in. When you're in, there are tournaments going on. A guy came up to me and said the big tournament was about to start in 15 minutes; it was my last chance to place bets. I don't know anything about betting on someone who plays video games; I have no idea who is good or who isn't. I asked more about how the tournament and how bets work. The guy said how they do it is card in the hat bets for each tournament.

So, each player in the tournament writes their name on a card and puts it in a hat, and the bettors involved pull a name out, and that's the person you're stuck with. The player pool was small. There were only eight people in the big tournament. So, only eight people can get in on the betting; when I was talking to him, there was one spot left. The bet entry fee was $625, and if my person won, I would win $5,000. It's not usually my type of entertainment to watch people play video games, but I was interested in trying something new. $625 is nothing to me if I lose it. I paid the fee and pulled the last card. The name on it said Steph. I said, 'Which one is he?' The guy laughed and said, 'She's over there.' She was the only female in the tournament. She played with the Milwaukee Bucks at the tournament. Every time she was blowing out her opponents, she won the tournament. It was the easiest $5,000 I ever made.

There's something about sports betting that gives you an adrenaline rush when you place a big bet. I've heard about Floyd Mayweather dropping big bets and Mattress Mack doing large outlandish bets. I'm not on their level with money to burn. But I really wanted to drop a million-dollar bet on something just to see if I could win. Parlays are also something that gives you an adrenaline rush. It's a lot cheaper than dropping a million dollars on a bet. On average, I was making some decent money when I did parlay bets. I'll do prop bets sometimes, but it's not really worth it.

One of the things about me is that I get bored fast, and I'm always looking for something new that gets me excited. That's why I bet on the NBA 2K tournament. I was consistently making money with the betting and fantasy sports I was doing, but I still wanted more. Something that was high-stakes and unique. That's when I turned to the underground betting world. I took a trip to Venezuela, and while I was there, I went to cockfighting events. Yes, it can be cruel for the roosters, and it's something I shouldn't bet on or support. But betting on those fights gave me the adrenaline I'm always looking for. It's exciting stuff. I made $6,600 in U.S. dollars. I was betting with drug smugglers and shady police. They were the only ones who had money to make big-money bets.

Going to different underground events, I would get introduced to various people and invited to other events. I went over to Saudi Arabia and never even heard of camel racing. But I was asked if I wanted in on the bets. I finally made my largest bet; it was $25,000 U.S. It would pay out $100,000 if I won. The people I was with were serious bettors. They had money I could only dream of. The only reason I could be around them was because my friend is a relative and I was with him. I didn't want to look like a lightweight, so I had to place a large bet. The camel I bet on won. I couldn't believe it. There was a dispute about whether it really won, and they didn't want to honor the bet. I said I wanted excitement, and that definitely did it with a mix of being nervous as hell. They agreed to give me $75,000. I wasn't going to argue with them over there. At least I still made a huge profit. I had enough excitement after that; mingling in the underground circles is kind of scary. After that, I made several other large bets in other sports with legitimate sportsbooks. I did well on those. Since becoming a professional gambler, I've added $1.3 million to my net worth with all the things I have done so far. My goal is to get around $5 million for my retirement fund.

Lauren

I worked at a store where lottery is available. I never played the lottery; I didn't have extra money to lose, but I was able to observe the way people played. Sometimes it's exhausting when someone is playing like 50 tickets. There were people who just bought scratch-off tickets. Sometimes it was a couple, other times it was like twenty. When I pulled out the scratch-off tickets, I would always look at 'em, one of the scratch-off games had a different look on the tickets sometimes. You could barely tell, you almost need superhuman vision, but certain tickets were a little darker than the others on the upper right side. When people would cash in the winners, I noticed the darker ones were always the winners.

At first, I thought maybe I'm imagining it. Then I started examining the difference between the winning ones and the losing ones on this specific game. There was no doubt a slight difference, but 99.9% of people wouldn't even notice it. You almost need a magnifying glass and a bright ass light to even see it. I thought maybe this batch of tickets just had rare design issues and our store got them.

When I was at a different store, I saw the same tickets in the display, and I decided to try my luck and test my theory. I bought 13 tickets, because Taylor Swift is my favorite artist. I looked at all 13 tickets when I got in the car, I saw three that I thought looked a little darker than the others on the upper right side. I scratched those off first. On the first ticket I won $100, the second one I won $5,000, and on the third ticket I won $20. I looked at the $5,000 ticket like 10 times, I was in disbelief that I actually won that much. I scratched off the other tickets, one of them I won another free ticket. The others were losers.

I went back inside and bought all the rest of the tickets for that game. I was looking at them as hard as I could to see if I could notice any darker parts like on the other tickets. I looked at all 18 remaining tickets repeatedly. I assumed they were going to be losers since I didn't notice any darker spots. I scratched all of them off, two of them I won $5. I still couldn't believe that I saw darker spots on the tickets, I thought my eyes and mind were seeing what I wanted to believe.

I asked my mom if she could see anything different. She couldn't tell. I asked my sister, she said she wasn't sure. I asked my dad; he said no way they would allow faulty tickets to be purchased, so he didn't believe it. I asked them all to buy as many tickets as possible of that game to see if they would win. I also bought some more tickets, on one of them I won $50,000.

My mom won $10,000 on one of her tickets. On the ticket my mom won $10,000, my dad finally agreed that he could see a darker spot on the right side of the ticket. But only because he was specifically looking for it. I don't think anyone would see it if you're not staring at it hard. We kept buying more of those tickets for another month, then suddenly we stopped seeing them. We looked on the website and it was no longer available. I don't know if they finally spotted the clue about the tickets, but I had to find a new game, because I was making money winning scratch-offs and I wasn't ready to quit. My dad said, that was a seasonal holiday game. So, maybe they didn't catch on and we should probably look for more of those types of tickets.

I would buy different types of scratch-offs, and of course I was looking hard to see if there were differences like the other ones. Unfortunately, all these tickets were the same, no clues that I could see. I would scratch them off, I didn't win anything more than $20, and most tickets were losers. I still worked at the store and handled the lottery throughout my shift. I listened to the different numbers people played when I entered it in. We also had to write the winning numbers down on this chart after they came out so people could see.

I'm an observant person, so I noticed in Pick 5 the first number had some of the same numbers frequently. 0, 4, and 2, seem like those were on rotation as the first numbers every few days. Some of the same second numbers would follow the first numbers. Like 03, or 41 would be the first two numbers the next day or sometimes every 5 days. I found my next lottery game with clues. I would print out all the numbers before I left work, looked at the numbers daily over and over trying to see if I could predict which two first numbers it would be in exact order. If I could get the exact order, I would win $25-$50 each time. The games are really cheap, it's 50 cents or $1. I didn't expect to get it exactly every day, but I was hoping to win a couple times a week at least. Just extra money I could use towards bills. I wanted to see if I could predict the first two numbers, then start working on figuring out all 5 numbers and win up to $50,000. I've won $50,000 before, maybe I could do it again.

Sometimes I would get the first numbers. Sometimes I switched it up and tried the last two numbers. I was mixing and matching. You could see there are numbers that would come out often, then it would flip the order. Like it would be 46, then 64. If I could pick up on the orders whether it was the first two or the last two numbers, then I had a quality strategy. It was hard predicting the numbers in the exact order. I started trying all 5 numbers box, hoping I could win that way, and I would sometimes.

The most I won in Pick 5 was around $25,000. Once again, I got my family involved, we sat around the table looking at the numbers that came out throughout the week, trying to figure out if we could predict the numbers. We put our money together and played various numbers we thought might win. We boxed some, did some straight, did the first two and last two exact numbers. When we worked together, winning started to happen more often. My parents were already retired, and my sister and I finally quit our crappy jobs, so we could focus more on gambling. At the store I was making less than $1,600 a month, sometimes gambling I made that in a week. We played additional games and went "all-in" with gambling becoming our full-time "job." The decision was worth it.

Cameron

When I was a kid, I would watch my dad and his friends sit around the table playing poker. Most times, they were so into playing that they didn't even notice I was there. I would walk around the table, looking at each person's cards. I also looked at their faces; sometimes someone would have an annoyed look on their face. Others would try to keep a straight face, trying not to show their excitement about the cards they had. My neighbor was having a garage sale selling poker items. I used my allowance to buy it. I told him it was going to be a gift for a family member, but really, I was keeping it for myself to play with my friends. We would play in my treehouse, practicing how to play poker. Our currency was bubble gum and candy. In high school, we would play poker at parties with a $10 buy-in, usually with five players. I was getting good enough to make around $500 a month. In college, I started playing online poker, No-Limit Texas Hold'em. I was addicted to playing it so much that I started failing my classes at UNLV.

My friends and I would go to poker rooms with a $50 buy-in. I was taking massive hits and constantly losing. I was visibly getting frustrated. A veteran poker player pulled me to the side and said if I wanted to start winning money, I needed to do low-stakes cash games or micro-stakes games. Learn how to win those before trying to play high-stakes games against people who have money to lose but usually don't lose. There was a difference between playing against high schoolers and people who have been playing half their lives. I started playing games for less than $5. Online, I would play thousands of hands a day. Eventually, I did well enough to make around $650 a month. My first year grinding with online games, I made around $8,000. It was enough to pay most of my bills. I got a refund check for $4,500. Since I had some extra money, I went back to trying to $50 buy-in games. I was getting better and winning money. On average, I was making like $3,200 a month, and that was just when I had the time to do it. I joined some high-stakes poker tournaments; sometimes I did well, sometimes I was terrible. Tournaments are crowded, long hours, lots of tables, and stressful; sometimes it doesn't even feel worth it. Some guys wear dark glasses, some guy's smell, and a lot of them are jerks, especially when they lose. I quit the job I was doing full-time and became a professional poker player. I was committed to becoming a successful poker player; it's been my passion since I was 13 years old.

The first tournament I played in since I quit my job, I won $7,500, finishing in the top 5. Normally, I'm a laid-back player; I just focus on trying to win. That win boosted my ego, and I started to talk shit to the other players in the next tournament. In that tournament, I won $12,000, getting 9th place. I feel like the key to me doing well was remembering what I did as a kid watching my dad and his friends. Observing people and reading their faces, watching their hands and arm movements, if someone hesitated or double-checks the cards. Using any little tell to my advantage. Most players will say they're doing that, but it takes a special person to know when someone is going to hold 'em, fold 'em, or whatever you think is going on in the moment. Doing acting on your part can be an advantage as well. I like to move my head around, like I'm cracking my neck and make it look like I'm making the biggest decision of my life sometimes. When really, I'm not that worried. You just have to make it look convincing and do it at the right time.

It was my dream to enter the World Series of Poker, and I was finally able to do it. It was the biggest challenge of my life. People on the circuit are the real deal. It's a humbling experience. When you get your ass handed to you, it deflates your ego and puts you back in your place. I was better in big-game tournaments with a smaller pool of players. When you're a professional gambler, you better be smart enough to know what works well for you and stick to it. Know your strengths and weaknesses. You're not going to win every time; you just need consistency. I've been doing this long enough now that I've been able to earn at least $80,000 a year full-time. I'll never reach Phil Ivey's level of success, but I'm still living the dream.

Walter

I teach mathematics, and one of my students asked me my thoughts on the probability of winning the lottery with an auto-pick. I didn't really know the answer. The lottery is something I've never studied. Some of my colleagues who are mathematicians have studied the subject. My brother is a software developer. I was wondering if all of us got together and put our knowledge and expertise together, maybe we could develop a system to predict winning lottery numbers. I'm not the first to think of the idea, and there's probably plenty of people who have created their own version of number predicting. I talked to my brother Terry; he was on board. I talked with a colleague, Glen, who has previously studied the lottery. I also talked with a tech specialist in AI named Connie and a machine learning engineer named Hong to help with the development and prediction. We agreed to work on this project together. We spent a lot of time working in Hong's lab putting together the prediction program.

After we gathered as much information as possible about the Powerball and Mega Millions, we figured out an algorithmic formula. There was enough data in our system to start going through millions of scenarios to try to generate as many accurate numbers as possible. Glen said it's important to figure out the lottery pattern daily, weekly, and monthly. So, Connie and Hong had to work on something that would work best for applied pattern recognition.

During testing, we wanted to keep it simple to see if we were on the right track. So, we started by trying to predict the Mega Ball for the Mega Millions and the red Powerball for the Powerball. The final number on those lottery games. The information was synced with those games, so when the numbers came out, the data was already in the system. As soon as the numbers came out, our system started running through the scenarios to try to predict the best pattern of numbers that might come out in the next lottery.

The Powerball numbers come out on Monday, Wednesday, and Saturday. The Mega Millions numbers come out Tuesday and Friday. So, there's days in between for our system to run through as many scenarios as possible before reaching the conclusion on what the numbers will be for the next drawing. In order for it to work, it had to be within the same week to get the best read on potential patterns. Going too far in advance gives inconclusive results. So, trying to run our predictions in June for numbers that come out in August would be a waste. I doubt the actual lottery system knows that far in advance.

On our first trial run, we start the predictions for the red Powerball on Sunday, so it would predict Monday's drawing. It was predicted that the number would be 16. I went to the store to play it. Picking the other five numbers right wasn't important to me at that time, so I filled in random numbers. I was only focused on 16 for the final number on the card. When the numbers came out, 16 was the correct number. I spent $2 playing it and won $4. On our first try, it seemed that our system was working.

We continued with running analysis for Wednesday's drawing. Once again, it was predicted that 16 would be the red Powerball number. I'll admit that I was skeptical that it was right. I thought there might be some glitch that keeps giving us the same final number. But I'm not a tech person, so I had to trust the system. I played 16 again, and it came out in the next drawing. It was a small sample size, but we felt like we were on to something, at least with the red Powerball. Glen said that once we start trying to figure out the other five numbers, that's probably where we'll be less accurate. We started running tests to predict all six numbers for Saturday's drawing. The best scenario in our system was 5-11-15-38-69, and red Powerball 14. What came out were 2-15-38-54-65, and 11. We were able to get two numbers correct.

Connie and Hong worked on ways to improve the information for prediction. They added the feature of the most consistent number coming out recently. It was 15. Every time we had a final conclusion on the predictor, I played the numbers. We would get a few numbers, so we were able to double the money each time.

One day, Connie and Glen figured out a way to increase the odds of getting more numbers right. I won't get into detail about that; it's proprietary information. When that feature was added, the predictor for the Mega Ball was correct 72% of the time. The predictor for the red Powerball was correct 69% of the time. It also predicted at least 3 numbers correctly on Mega Millions, 76% of the time, and 71% of the time for the Powerball. We were making money with the predictions. It was clear our system was working. We still weren't able to get all six matching numbers. One of our predictions had five numbers correct, and had some other things correct and we won $2 million in total winnings. During that time, we wanted to try an expanded experiment and we bought hundreds of tickets at different locations, and played all the numbers with the best percentage to come out.

When that happened, we felt like it was only a matter of time before we got the jackpot correct. I left my job in teaching to focus more on our lottery project. I wanted us to expand to state lotteries as well and try to win as much money as possible. We collected information for lotteries in the surrounding states and were working to see if our prediction system would work for other games that had fewer numbers. Our system was good at predicting five numbers for one of the lotteries, so we won six figures several times.

We had conversations about whether to keep the program for ourselves, license it, or sell it to a large tech company. Hong reached out to a cousin in China just to see what type of offer we could get for selling our software. There was a company willing to buy it for $22 million. $10 million in cash and $12 million in stock. Glen said if we won the Powerball, we could make $222 million and get to keep it. That's the smarter option. We all agreed that none of us were hurting for money, so there was no need to desperately sell it off. We've been using our program for two years, and there's never been a time we lost money. There are many times where we broke even, and we still haven't won the Powerball or Mega Millions jackpot yet, but we're all millionaires now, thanks to winning various times in other games.

Angelique

My father was a racehorse trainer. So, I spent a lot of time at the racetrack. I loved watching the Kentucky Derby; everything was interesting to me. The outfits, the horse names, the betting, the races. It made me want to know more about the industry. I would ask my dad every question I could think of about horse racing. When my family would go to the racetrack, my parents would let me pick a horse to bet on with my allowance. My dad told me the goal should be to make a profit with my money. Don't just pick a horse based on its name, the color of the horse, or because it's the favorite. Value and profit are what matter. There are various factors to focus on when betting on horses. What are the conditions of the track?

There are different types of racetracks. Weather conditions can play a factor. Knowing about different breeds of horses is very helpful. There are different race styles. Different breeds have different speeds when it comes to different events. Learning about the trainers is just as important as learning about the horses and racetracks. Trainers that are known for having success with horses usually know how to look out for the horse's best interest and well-being so they can make as much money as possible. Good trainers care about long-term success. Treating them well and making sure they're happy. Remember, horses have feelings and moods. Bad trainers care about short-term success. If they can run a horse into the ground immediately, trying to capitalize off their instant success, they'll just move on to the next and try to recreate another popular horse. So, if possible, try to learn about trainers, their reputations, and their training styles. Be aware of trainers who are known for having horses with failed drug tests.

After you learn about the trainer, start learning about the rider. Consistency is key for me. Jockeys that get top-two finishes regardless of the horse, level, or race conditions are obviously winners. It's also helpful when doing a top-two finish bet. How they bond with the horse in a race is an important partnership for a victory. A pitcher and catcher need to be on the same page, or a center and a quarterback. The same applies to a horse and rider. If the conditions are sloppy, how they're going to get through it together is a key factor in victory. Knowing a horse's and rider's history at different racetracks and levels they've raced at is helpful. Learning about different breeds and different events can all be key factors to know before betting.

It's easy to go to the racetrack and place a bet on the horses with the best odds to win. But if you're a professional horse gambler doing this to make a living, it takes a lot of work to study before making the bets. All these places give you odds, and people assume the favorite will usually win the race. But if you study the races thoroughly before placing bets, you'll learn that picking the favorites is statistically a bad assumption. It doesn't mean that they won't win, but most of the time they'll be "upset." Longshots can have higher payouts if they win. But if you're trying to win on a regular basis, longshots are not the best value plays. Remember, the goal is to win money consistently. I don't recommend it for everyone, but Dutch betting has done well for me. Backing multiple outcomes in one event to profit from one of my chosen outcomes winning, is a good strategy. When I've studied the different factors, like I previously mentioned, it helps me narrow down who might have the best chance to win. There might be a certain horse and rider that do well in certain conditions, but the trainer has a so-so reputation. It still might be worth a bet. Another horse can be on a hot streak; it could go either way if the streak continues, or it might come up short. Still, it might be worth a shot compared to the rest of the field. You might find out during your research that the longshot horse checks all the key factors you're looking for in a specific race. That could be worth betting on as well. Dutch gives you options for more winning. I don't mind taking risks; I can afford it at this point.

I benefit from my father's experience as a horse-race trainer. He has a lot of trusted friends in the industry. So, I'm able to learn about jockeys, trainers, horses, racetracks, and other valuable information to apply to my factors when I decide to place bets. Too many people rely on other people's information and strategies. Develop your own and do what works for you. There are people who are hardcore believers who believe that knowing about trainers, jockeys, and horses doesn't matter. When it's race time, what will happen will happen. There is truth to that. But from my experience, the more valuable information you're able to use, the better. It's how I got rich. I couldn't imagine doing it any other way. Learning about different horses is no different than learning about humans, believe it or not. If you know certain horses like to stay in the middle of the pack most of the race and one of them tends to break away late, that's a good thing to know. If you watch track sprinters, there are some that start out the blocks slowly and can turn on the jets at the end. Or some people get off to a fast start but constantly end up in 4th place. You're familiar with their style and outcome.

Learn about who is likely to lead, trail, or stay in the middle of the pack. Learning about critical factors should be important to serious bettors, in my opinion. Some of the things I do as well. Taking meticulous notes. I use every piece of information I can keep track of. Whether it's for horses I like to bet on or potential horses I'm interested in, that goes on the pros side. The con side has horses to absolutely stay away from. They could be trending down because of a losing streak, maybe from bad treatment from a trainer and owner wearing them down. How they act before a race can be a pro or con. They might struggle on certain surfaces, or whatever negatives and red flags are on your radar. Horses have stats and noticeable vibes too. Use that to your advantage.

This might sound like a commonsense thing, but most people don't focus on it as much as they should, especially when they're starting out. Learn about different types of bets and odds. I'll repeat it again. The goal is to make money with value. So, understanding betting is important. Do not bet if you don't understand handicapping. Standard wagers and Exotic wagers. Win bets, Place bets, Show bets, Exacta bets, Trifecta bets, Superfecta bets, and Quinella bets. Whatever it is, learn about it. Learn and study odds; this is ultimately how you win a lot of money or lose a lot of money. So, for example, if a horse has 70/1 odds and you bet $10, You will win $700. I have a friend who went to the racetrack, just making bets, and didn't understand how odds work. Know the difference between 70/1 and 5-2 odds, or whatever is on the board. Know what the payouts are.

Do things that give you the best chance to win money. Betting on horses is very unpredictable; there are trends you can follow and information you can use to make some money consistently. But I don't know anyone who gets it right every time. You're going to take losses. Even with all the knowledge I had, it took years of losses and learning what worked best for me to make money consistently. I'm still learning daily, and I've been doing this for over 20 years. The math and probabilities of betting on horses never stops. Things always change, and you need to understand how to adapt and change as well. That's how you stay successful in this industry. My dad always said, "Smart people learn from errors and experience. Idiots are arrogant and ignore learning."

Aidan

I was out doing errands with my girlfriend. She said that while we're out, she wants to play the lottery. She had never played before and thought it would be fun to try. She picked random numbers: the date of the day we started dating, the birthdays of people she knew, her favorite number, my favorite number, whatever she could think of. She tried a couple different games but got several tickets. She lost on most of them but two. On one of them, she won £1,750; on the other, she won £50. She was so happy, she jumped up and down like she had won £500,000. Naturally, she wanted to play again to see if the luck continued. The next time she played, she won £350. I couldn't believe her luck. So, I had to start playing too; I was getting jealous. I bought like 10 tickets and didn't win anything. I joked with her and said I'm letting her pick all my numbers from now on. The next time I played, I won £10, but I didn't really win; I just got the money I spent back. I decided the lottery wasn't for me and left the playing to her.

I was on the phone with my mum, who lives in the United States. I told her about my girlfriend being on a winning streak with the lottery. She wanted her to tell her some numbers to play in the United States. My girlfriend grabs the phone with excitement and says, play this, play that, do this, do that. My mum played every number she told her to play. When my mum called the next day, she told me she had won $1,000. I was happy for her and jealous. These two were winning, and it made me want to try again. My mum sent $500 to us to thank us for the numbers, and of course she wanted more numbers to play. It became a regular activity for my mum and girlfriend to talk a few times a week about lottery numbers. It was like my girlfriend was a lottery savant, even though she was just picking whatever random numbers popped into her head. There was no real method to the madness. Nevertheless, what she did worked and was making us money. My mum won a few more times. My girlfriend won a couple more times. I even won once.

My mum started telling other family members and they wanted in on winning some extra money. My girlfriend really started taking it seriously when people offered to split it with her if she gave them the right numbers to play. My sister won $400, and my aunt won $630. I was in awe that Emily was picking all these right numbers.

She was just giving them three and four-number games. Nothing in which you can win millions of dollars. Emily wanted to win life-changing money. We went over to the United States to visit family, and while we were there, she wanted to play the lottery in the United States. She won $150 playing Keno. She bought a bunch of lottery tickets and won $500 on one of them. My mum went with her, and they picked some numbers together. They strategized when picking numbers. Emily noticed two of the same numbers came out together a couple times that week. My mum said sometimes three of the same numbers come out together and said, let's play all the same numbers. The numbers came out, and they won $500 and split it. Emily had $1,000 in winnings. Before we left, I went to play the lottery with my mum. I picked some numbers, and I ended up winning $166,000. Emily was the jealous one now. We wanted to have a fun competition with each other. From that point on, we kept track of wins and losses with winning tickets and money. The loser each week had to do something for the winner. Before we moved to the United States, she was up £660 to my £410. My luck was a lot better in the United States; I won big. I played the Powerball and won $50,000. I was crushing her in the competition in the United States. I bought her an engagement ring with the money.

One day, mum wanted to have a lottery party for our family. We all bought thousands of tickets. We had the fastest scratch-off, seeing who could scratch off tickets the fastest. Guess the amount for winning tickets; the closest number wins the tickets. There were other games as well. The party was fun, and it also gave people in our family some extra money. If lucky, maybe even life-changing money. It was over $16,000 in winning tickets. I said we should do this every week. So, we bought at least a hundred tickets every weekend. Some of us also put money together to buy lottery tickets. We split money on the winnings. Our pool of money with weekly winnings was more than what each person was making at our job's weekly. That's when we decided we were all going to quit our jobs and continue to buy as many scratch-offs and lottery tickets as possible. One of the winning tickets was $100,000. Plus, there were other winners with smaller amounts. That week, my cut was $15,000. Our family was able to elevate each other by putting our money together and taking a chance at winning the lottery with as many tickets as possible. We bought a restaurant as well just in case we didn't make enough money to continue to invest in the lottery. So far, the ROI on buying tickets has outweighed the losses. I would recommend every family invest in the lottery together, even if it's just $5 each. Take the burden off spending a bunch of money yourself. All it takes is just one ticket to go from being in debt to being rich. Just make sure you don't put yourself in further debt trying to win big.

Donna

I moved to Las Vegas for a job opportunity. I was there for 7 months when the company decided to close and move operations to India. I didn't qualify for unemployment, so I had to figure out what to do. I really liked Las Vegas and didn't want to move back to Spearfish, South Dakota. I was looking for other jobs in the classifieds, but nothing was available in my field, where I got my degree. When I saw a flyer for a chance to win $10,000 playing in bingo tournaments for a $200 entry fee, it was a chance I was willing to take. It had been a few years since I played bingo, but it's easy to play. I entered the tournament, and there were hundreds of people in it. 1st place wins $10,000, 2nd place wins $7,500, 3rd place wins $5,000, 4th place wins $2,500, and 5th place wins $1,000. Sixth place to tenth place wins $500. They had two-day tournaments on a weekend almost every month. I wasn't expecting bingo to replace my full-time job; I was just hoping to make some extra money to help me survive until I found another job.

In my first tournament, I won 4th place. The next month, I entered again and was in eighth place. I spent $400 to enter and won $3,000 in my first two tournaments. As long as I kept winning, I was going to keep playing. The next month, I got ninth place. I had to take a month off after that due to a scheduling conflict. But when I returned to play again, I got sixth place. The last tournament of the year, my streak ended; I didn't have anything good. A $3,000 profit isn't bad for just playing bingo.

When I wasn't playing bingo, I spent time in the casino mostly playing the slot machines. I wasn't winning much money, but it's addicting to think that the next one will be the jackpot. It took a lot of playing, but eventually I won $2,600. Another time I won big was for $750. I won around $100 playing bingo at small bingo venues. I was waiting anxiously for another large bingo tournament to start. I entered another tournament where the jackpot was $5,000 for first place. 2nd place was $4,000, 3rd place was $3,000, 4th place was $2,000, and 5th place was $1,000. Unfortunately, I didn't get into the top 5.

Since I started gambling in Vegas, I had won over $14,000. Even with the losses, I still had a great profit return. When I wasn't gambling, I worked part-time. I finally had to make the decision to move to another city to get back to working full-time in my field of work. I landed a job in Costa Mesa, California. They don't have a lottery in Vegas, so I was excited about being able to play the lottery in California. On one of the tickets I bought, I won $50,000. That was almost the salary at my job. After taxes, I put most of it in savings and invested in some stocks. The rest I was going to use towards gambling to see if I could increase my winnings even more. I traveled to Vegas when I heard about another large bingo tournament. I won $4,500. While I was there, I gambled at the casino as well and won over $6,000. My weekend trip was a $10,000 success.

When I got back to Costa Mesa, I had this disappointed feeling. Going to work at a 9-to-5 job was so unfulfilling. Every time I gamble, it's fun and exciting. Even when I lose large amounts of money, I still enjoy the experience. I wanted to have that feeling all the time. That's when I decided to take the risk and quit my job. I wanted to try to become a full-time gambler. I was only making around $62,000 a year at my job. If I could get close to that, I would be fine. I laid out a plan for the gambling routine I was going to do. I started playing online gambling throughout the week for at least 5 to 6 hours a day. I kept playing various games, trying to get better. I also started spending $100 a week on lottery tickets. $60 on scratch-offs and $40 on lottery tickets.

On the weekends, I went to Arizona to play at the casino. Sometimes I would go to Vegas. This became my regular routine. Some weeks, I would lose more than I spent. Sometimes I would make an unbelievable amount of money. In my first year of gambling full-time, I made about $43,000. So, I quit my job to make $20,000 less. But I was a lot happier doing it. My second year, I made over $58,000 gambling full-time. In my third year, I made close to $400,000. I felt validated that I had made the right choice. The year after that, I took a significant hit and made around $52,000 gambling. The more I was making, the bigger the risk I took. If I would consistently play it safe, my losses wouldn't be that bad, but my biggest struggle was not getting carried away with reckless spending. I would travel to casinos across the country, racking up expenses. If I stuck to the plan, staying close to home, my finances would be better. I was making a profit, so I wanted to live a little. Having a millionaire mentality when I'm not a millionaire is silly. I had to learn that lesson.

I'm in my seventh year of gambling full-time. I've made over $700,000. According to my calculations, I'm making at least $250,000 more than I would've made had I stayed at my job with slight raises. My goal is to get at least a million dollars in winnings. I probably would've reached that goal if I hadn't made dumb decisions when I was winning big. I would love to win $600 million with the Powerball. But I'm realistic; that will probably never happen. I'm fortunate enough to make as much as I have. It was a risk to quit my job to gamble. Several people thought I was stupid to make that decision. When I first did it, my parents said I wasted their money and student loans going to college if I wasn't going to have a career in my field. I didn't enjoy my career and switched to something I did enjoy. I paid my parents back and paid off my student loans. I'm not sure I would've been able to do that without gambling.

Sylvester

When I played the lottery, I was just playing a bunch of different numbers, hoping I hit something. It wasn't until I met you that I realized that I was doing the lottery wrong. I was losing $50 a day trying to win. After our conversation, I changed the way I played the lottery. I focused on trying to learn one game and master it. I wrote down the numbers every day and studied the numbers that came out. Like you said, all you have to do is take a minute to write down the numbers and learn the numbers and patterns. You'll quickly see there's something there; you won't truly understand it for a while. You'll at least see there's potential to figure out the numbers if you stick with it. I took a break from playing and waited 3 months until I started playing again.

When I felt like I had an understanding of the pattern with certain numbers, I was ready to start playing again. I took it number by number. One month I focused on the number I saw the most, which was 8 at the time. Then I focused on the number I saw with 8 the most, which was 7. I also noticed 8 went with 5 often and 7 often. I won six times that month because I was able to figure out various combinations of 8, 7, and 5. Instead of losing $350 a month like I used to. Now I had won over $500 that month, just by closely paying attention to the lottery.

The more I followed the lottery, the more I picked up on understanding patterns. It's obvious there are numbers that go together more than others. All you have to do is look at the numbers daily if you keep track. The next month, I won $1,249, but I spent less than $30. I stayed patient, not trying to get rich overnight. I wasn't spending $50 a day and was focused on making a profit. It was hard not to spend $20 playing various numbers. I had to pick a combination, and I had faith I picked the right one.

The previous way I played; I won maybe 10 times a year. Now I'm winning at least 4 to 8 times a month. It's been a great supplemental income. A lot of people are always looking to make extra money without adding a bunch of hours with another job. Playing the Pick 3 is a viable option if you're willing to be patient and learn how to play to win. I'm still focused on learning Pick 3 and winning more before I move on to trying bigger games. I would like to learn all of them someday. For now, my goal is to win at least 15 times a month with Pick 3. I appreciate you sharing your knowledge with me and putting me on the right path to hopefully becoming a successful professional gambler.

Martin

I've been gambling since I was 13 years old. I had a betting pool at my school for NFL games. We played an NFL pick'em, picking all the games weekly with weekly payouts. I did it all the way through high school, sometimes making $400–$500 a week. I had jobs cutting grass, raking leaves, and shoveling snow. All the money I made with pick'em and betting in school I never spent. I had like $70,000 to $75,000 saved up at 19 years old. I could've used it to go to college or start a business, but of course I wanted to use that money for gambling. I couldn't wait to turn 21 years old; that's when I could go all in with gambling.

For two years, I watched every YouTube video possible from people who gambled full-time. I read several books about gambling. I read articles and Wikipedia pages on gamblers who won a lot of money and lost a lot of money. This was my college education, as far as I was concerned. I soaked up as much information as possible. I was going to do what it takes to win millions of dollars as a professional gambler. When I was 21, I set up accounts with four sportsbooks. I still wanted to learn more about arbitrage betting before doing it. I had a notebook on gambling experts, keeping track of how well they did for two years.

Every key point they mentioned about why it was or wasn't a good idea to place certain bets. If they had a track record of doing well with their analysis and suggestions, I had them in a column to use when I wanted to make wagers. I had another column for the people that were full of shit; their analysis was trash, and most of their suggestions turned out wrong. You'll lose money listening to those morons. I practiced manipulating lines, seeing how well I would do with mock bets. I kept track of it for two years. I learned a lot by doing it. It would have been nice to actually bet, because I did pretty well and would've won a nice chunk of money during that two-year period. But at least I knew I had the potential to do well because I paid close attention and studied my ass off to learn the right trends. Learning a lot so you can earn a lot was my motto.

You probably have to spend a minimum of 8 hours just learning and studying. Just like an analyst is preparing for the game he's covering; a sports bettor also has to do that. I have a friend who works in sports production with TV trucks for broadcasts. He was able to send me the All-22 film for football games if I needed it. This helped with my fantasy football teams as well. Understanding the teams, coaches, and players is an important factor in sports betting. Most people don't have time to do this. But if you're gambling full-time, it's basically your job. Paying close attention to injury reports at all times is another important factor.

An injury to a key player can make or break your bet. If there's smoke, there's fire. For example, there were reports that a quarterback didn't throw all week. His shoulder was banged up from the previous week. The team wanted to save him for game day. He had every intention of playing, and the team said he'd be ready. You could tell when he was hit in that shoulder, no matter how much rest he got that week or being shot up with whatever. He's not going to be effective, whether he plays or not. I stayed away from betting on that team because the backup sucks and so does his offensive line. I also benched him in fantasy. He didn't end up playing; he couldn't throw the ball further than 7 yards. Closely monitoring the situation was very valuable for me. Other people were confident he was playing because it was reported he'd be good to go. But the clues and facts indicated that it wasn't even close to happening. Many people took a hit by believing the reports.

One of the things I picked up from learning about sports betting is that coaches are just as important as the players. You're probably thinking, 'DUH, DUDE!' But what I mean by that is from a strategic standpoint. In football, for example, coaches are going against each other like the players. An offensive coordinator vs. a defensive coordinator. Focus on the matchup and see who has the advantage.

If you have a crappy offensive coordinator like Matt Canada going against Bill Belichick's defensive scheme, smart money takes Bill to win that matchup. That doesn't necessarily mean that the Patriots will win. But I would put that on my chart in the positive factor section. Or if you have a crappy defensive coordinator like Alex Grinch, go against Ryan Day's offense. Smart money takes Ryan to win that matchup. Even in college, knowing about the GA's and experienced offensive and defensive analysts might be a helpful advantage. Knowing all the personnel involved is a critical factor in winning the game. With odds, always watching the way lines move or don't move is important. Don't take the suckers bet. I had to learn that. Sportsbooks, casinos, and whoever is in charge of payouts don't want to see you win. So, why would you trust them? That's why manipulating the lines can help you win money if you know what you're doing.

If a team is flying from Seattle, Washington, to Miami, Florida, that's information to use in factoring a bet. If the weather is bad, and a player is known to fumble a lot, regardless of the weather. That's information to use. In the NBA, it's known that visiting teams that go to Houston and Miami party a little too hard and can be off their game sometimes. That's something to think about. I'm always trying to acquire as much information as possible. If you're doing this full-time, that's what you should be doing. I put in 8–12 hours of studying a day sometimes, just preparing for upcoming bets.

There are also people who want to gamble on everything; I personally don't feel like that's wise. I believe in quality over quantity. I'm trying to get the most bang for the buck. I hate baseball; it's boring, and it's a hard sport for me to bet on. Especially this new-age MLB stuff—shifting, pulling a pitcher after one at bat. It's a bunch of ridiculous bullshit, in my opinion. I like to prep for my bets, and baseball is too unpredictable for me, so I stay away from it. You don't have to bet everything. I'd rather stick to something attainable than go for some wild 10-team parlay. Yeah, it's nice if you hit it all, but there are better things you can do and win just as much money if you know what you're doing and know the tricks of the trade.

I'm trying to think of all the best tips I can share with you. Another thing I learned that's valuable is line shopping. If you're new to sports betting, you might be prone to going with the first thing you see. But it's better to take your time to find the best bang for the buck. Tracking lines and sportsbooks are important. Things can constantly change, like stocks. You want to get in on something where you maximize your profit. For each sport and league, I bet on, I have a different checkpoint system.

So, with college football, I have like 20 things I need to go over before I make a bet. If at least 13 positive things don't stand out, I won't make the bet. Having a detailed database has made me a lot of money. That's probably extreme for most people, but it works for me. There's so much to learn about handicapping. Most veteran gamblers know that whether it's sports betting or cards, constantly studying and gaining experience are the best things you can do. Learn, learn, learn your craft. Take your time and study. It takes some time, so practice a lot. Put yourself in a position to succeed. Yes, I sing that to myself all the time. It keeps me focused on the task at hand. Some gamblers get complacent and forget that their mission is to win more than they spend.

I've been studying this for a long time now. I can't prepare for everything; at some point, unexpected things will happen. I've won, I've lost, and I've broken even. I've been smart with my betting decisions. I've been stupid with my betting decisions. I've learned from my mistakes. I've developed bad habits that are hard to break. I was financially able to get into gambling full-time the moment I became an adult and have been doing it ever since. I was able to take time to study it for two years full-time before I officially started betting on games. I say all this because I want people to understand that if you're going to do this and you want to be successful at it, you have to put in a lot of time and effort. I've seen lots of people go broke, declare bankruptcy, lose families, and get severely beaten up because they couldn't pay back money. I've also seen people go from being broke to being rich beyond their wildest dreams. Things can change at any moment with gambling, good and bad. Be prepared for either one to happen. Have fun but take it seriously as well. Gambling isn't for everyone; it's important to be self-aware and know whether you're good at it or just an addict. When you realize you're not good at it, quit ASAP!

Ray

My gambling experience has been a rollercoaster. When I first started gambling, I had \$28. I used \$16 for food, \$4.50 for a bus pass, and the rest for the lottery. I figured I had nothing to lose and a lot to gain if I won. I bought a \$5 scratch-off ticket and a \$2 Powerball ticket. I won \$100 on the scratch-off ticket and \$100 with the Powerball ticket. At my temp job, I barely made \$200 a week, so I was beyond excited to make that much in a day without working long hours in a hot, smelly warehouse. After bills and child support, I usually have less than \$40. I wanted to treat myself to a nice meal at a steakhouse. When I left, I found \$200 in the parking lot. This was my lucky week. I wanted to see how long this lucky streak would go. I was thinking about how I won \$200. I found \$200.

I guess 200 is my lucky number this week, so I need to play that in the lottery. I played 200 straight and boxed it. I kid you, not 200 came out, and I won \$583. It felt like I was dreaming; I had never been this lucky in my life. You know I had to keep it going, so I went to Hollywood Casino to see if I could win more. I stayed with the theme of \$200, so I spent \$200. Unfortunately, the streak ended, and I lost \$190. Had I walked away when I was up, the streak would've continued, but I got greedy. When I left the casino, I went to play the lottery. I bought 20 scratch-off tickets and 8 lottery tickets. I'm weird when it comes to signs and themes. I figured my luck started when I only had \$28 left, so maybe 28 is a lucky number for me. Again, I kid you not, I won \$2,020 dollars. Here's a photo of the ticket. For some reason, I win with 28 and 2's and 0's. It's crazy!

I was making more money playing the lottery than working at my temp jobs, so I wanted to see if I could find a viable way to make a living gambling. I started searching online to see if there were people who gambled as a career choice. There were a lot of things about professional gambling. I read over 50 articles, learning everything about the subject. I printed the ones of people who were successful at it, and I printed the ones of people who failed at it. I highlighted the critical information for both scenarios. I was excited about the potential career choice, but I also felt like I was being irrational, thinking I could actually do this full-time.

Gambling full-time is something for rich people, not someone with $1,026 in his savings account. But taking risks with gambling was paying off for me; in the worst-case scenario, I would go back to working temp jobs. The first step was identifying what I was going to do to win the money. Scratch-off tickets and lottery tickets were obviously at the top of my list. In most of the articles I read, the people who were most successful as professional gamblers played online games. So, I had to decide which games I had the best chance to win. I have a decent track record with blackjack, so that was going to be one of them. I also learned some countries don't deduct taxes if you win. I love that, so I kept that in mind for gambling possibilities. There's like 70 countries that offer online gambling opportunities.

One of the key takeaways when I studied the articles was that if I'm going to make gambling a job, treat it like a job. When I worked at the warehouse, I worked from 3 to 11 p.m. five days a week. I needed to make a set schedule for gambling as well. Giving yourself breaks is important, so you don't burn out. We all need mental and physical breaks. Sitting in front of a computer for 10 hours a day is harmful. Have a work-life balance. Another key thing I highlighted was knowing when to quit. That can be a challenge. I had a hard time doing that when I went to the casino. When you win, you keep thinking, 'Maybe I can win more!' If you lose, you keep thinking, 'Maybe I can win it all back with a couple more tries!'

One of the tough lessons early on was making sure I find online gambling sites that are legitimate. I gambled on a site that was illegal, and I didn't know it. When I thought I won thousands of dollars, it turned out to be false, and I lost hundreds of dollars that I spent trying to win the money. That mistake set me back. I almost quit; I thought that might be a sign to get out now. But I kept going, thinking about the bigger picture. Professional gambling takes persistence; you're going to have pitfalls. You have to keep reminding yourself of that. I played different games online at home.

I played the lottery, bought scratch-offs, and sometimes went to internet cafes. I was losing money. I won a sweepstakes prize at the internet cafe; it was a brand-new laptop computer. It was worth like $1,300. I needed money, so I sold it for $1,400 online. I paid some bills, then I bought some scratch-off tickets. One of the tickets was a $5,000 winner, and another was a $1,000 winner. I had a couple that were like $2 winners. When I cashed the tickets, I decided I wanted to take a trip to Vegas and gamble there.

I had never done sports betting before, but I wanted to try it. I made several parlay bets. Some money line bets and point spreads, I ended up winning over $12,000. I gambled at a few casinos. I spent about $600 and won over $1,600. My week in Vegas was successful; I won over $17,000. I couldn't believe I was really making a living gambling. It wasn't all good, though. I fell into the other temptations of Vegas.

I hooked up with a few prostitutes. There was one named Black Bunny; something about her made my wood hard for hours. We had sex four times in one night. I was ready to go again, 5 to 10 minutes after we were done each time. I know the dangers of hooking up with hookers, but I couldn't help it. I was horny, happy from winning a bunch of money, and I was also intoxicated with alcohol or drugs. My judgment meter was broken. She told me some sob stories, saying that she only does prostitution so she can pay for the medical bills for her terminally ill son. She needed to make $30,000 ASAP, and she was $5,000 away from what she needed.

So, my dumb ass offered the $5,000. Of course, she gladly took the money without blinking. The next day, after I finally sobered up,

I started doing the math on how much money I had left. During my intoxicated binge, I spent so much money that I only had around $4,800. I can't even tell you where all the money went; it's that much of a blur. I needed to get the hell out of Vegas ASAP.

I went back to Ohio. When I got back, I found out my son needed a medical procedure done that insurance didn't cover. It was going to cost $5,750. If I wasn't a reckless idiot in Vegas, I would've had the money for my son to get the procedure done. I honestly wanted to kill myself. I was that depressed about it. After all the money I had, I was down to $736. I needed to come up with $5,000 ASAP. I was hoping I would get another scratch-off worth $5,000 like before. I knew the odds were unlikely, but I felt like since I've been gambling, I defied the odds several times. I spent $100 on lottery tickets and scratch-offs. I won over $800. I felt a little better, but I was still stressed out about needing to win more.

I entered a euchre tournament, and the winner gets $1,500. It had been a while since I played it, but I paid the $50 entry fee. There were a lot of old people in it. I can't lie; my exact thought was that none of these old asses will beat us. I'm looking around; one guy could barely move his arms, and there was a lady who had a hard time seeing. It almost felt wrong to play against them, but I desperately needed money, so fuck them was my mentality. As expected, we won the tournament. I was halfway to my $5,000 goal. My ex sold her car for $3,800. So that took the pressure off trying to come up with more money. Our combined money paid for the surgery.

I was back to where I started, down to $41. I hadn't been that broke since the $28. Once again, I had that nothing-to-lose mentality. I spent $20 on scratch-off and lottery tickets. I scratched 'em and double-checked 'em, hoping I didn't miss something. I didn't win anything. I was down to $11. I spent $2 on a Lucky for Life ticket. I won $200. It's unbelievable how many times 200 has saved my ass. A friend of mine asked me if I wanted to go to the casino. I was hesitant;

my casino experiences are usually my downfall. But I said, what the hell, I'll do it. I spent about $80 and left with $360. I had to go back and read the articles I printed. Focused on the highlighted stuff. I had lost my way and needed to get back on track. I was doing more of the don'ts and needed to stop making those mistakes. I got my mind right and laid out a plan again. I kept myself disciplined and consistent. I still have ups and downs as far as wins and losses, but nothing crazy like before. I make around $3,000 a month, mostly from online gambling full-time. I'm not ballin' like a lot of professional gamblers, but I have a comfortable life. It beats working in a hot, smelly warehouse for $9 an hour.

(List of My Books)
10 Categories of Trivia
200 Mixed Trivia Questions
AWARDS AND HONORS TRIVIA
Black Trivia Book
Colleges and Universities Trivia
College Trivia Nights
Healthy Habits
Health and Medical Trivia
I'll Tell You Why I Do It
Let's Talk About Women's Basketball
Making Sacrifices
Men Are Victims Too
Men in Sports Trivia
Music Trivia Book
People in Films and Shows Trivia
Real-life Lessons
Reflecting on My Interactions with Strangers
Religion Trivia
Stories of Struggling Fathers, Broken Families, and a Broken System
Successful Professional Gamblers
The Neighborhood Competition
Trivia About Occupations
Trivia About Politics
True or False: 5 Categories Edition
True or False: 2024 Olympics & Paralympics Edition
True or False: Organizing & Movements Edition
True or False: Powerchair Football & Power Soccer Edition
True or False: Video Games & esports Edition
Women in College Sports Trivia
Women in Sports Trivia
Worldwide Locations Trivia

www.ingramcontent.com/pod-product-compliance
Lightning Source LLC
Chambersburg PA
CBHW050554160726
48003CB00002B/890